MIX
Papier aus verantwortungsvollen Quellen
Paper from responsible sources
FSC® C105338

Isma Younes
Muhammad Shafiq
Fouzia Shafiq

Using Geographical Information Systems (GIS) to study the concentration of major air pollutants in Lahore City of Pakistan

Anchor Academic
Publishing

Younes, Isma, Shafiq, Muhammad, Shafiq, Fouzia: Using Geographical Information Systems (GIS) to study the concentration of major air pollutants in Lahore City of Pakistan, Hamburg, Anchor Academic Publishing 2015, Hamburg, Anchor Academic Publishing 2015

Buch-ISBN: 978-3-95489-404-8
PDF-eBook-ISBN: 978-3-95489-904-3
Druck/Herstellung: Anchor Academic Publishing, Hamburg, 2015

Bibliografische Information der Deutschen Nationalbibliothek:
Die Deutsche Nationalbibliothek verzeichnet diese Publikation in der Deutschen Nationalbibliografie; detaillierte bibliografische Daten sind im Internet über http://dnb.d-nb.de abrufbar.

Bibliographical Information of the German National Library:
The German National Library lists this publication in the German National Bibliography. Detailed bibliographic data can be found at: http://dnb.d-nb.de

All rights reserved. This publication may not be reproduced, stored in a retrieval system or transmitted, in any form or by any means, electronic, mechanical, photocopying, recording or otherwise, without the prior permission of the publishers.

Das Werk einschließlich aller seiner Teile ist urheberrechtlich geschützt. Jede Verwertung außerhalb der Grenzen des Urheberrechtsgesetzes ist ohne Zustimmung des Verlages unzulässig und strafbar. Dies gilt insbesondere für Vervielfältigungen, Übersetzungen, Mikroverfilmungen und die Einspeicherung und Bearbeitung in elektronischen Systemen.

Die Wiedergabe von Gebrauchsnamen, Handelsnamen, Warenbezeichnungen usw. in diesem Werk berechtigt auch ohne besondere Kennzeichnung nicht zu der Annahme, dass solche Namen im Sinne der Warenzeichen- und Markenschutz-Gesetzgebung als frei zu betrachten wären und daher von jedermann benutzt werden dürften.

Die Informationen in diesem Werk wurden mit Sorgfalt erarbeitet. Dennoch können Fehler nicht vollständig ausgeschlossen werden und die Diplomica Verlag GmbH, die Autoren oder Übersetzer übernehmen keine juristische Verantwortung oder irgendeine Haftung für evtl. verbliebene fehlerhafte Angaben und deren Folgen.

Alle Rechte vorbehalten

© Anchor Academic Publishing, Imprint der Diplomica Verlag GmbH
Hermannstal 119k, 22119 Hamburg
http://www.diplomica-verlag.de, Hamburg 2015
Printed in Germany

Dedication:

To our families

Left blank for pagination

Contents:

Dedication: 3

Contents: 5

Chapter 1: Introduction 7
    1.1   Global Effects: 8
    1.2   Air Pollution in Pakistan 10
    1.3   Air Pollution in Lahore City 11
    1.4   Aims and Objectives of the Study 12
    1.5   Data Sources 12
    1.6   Methodology 13

Chapter 2: Air Pollutants, Source and their Effects 17
    2.1   Introduction 17
    2.2   Types of Pollution 18
    2.3   Air Pollution 18
    2.4   Air Pollutants 21
    2.5   Categories of Air Pollutants 21
        2.5.1 Primary pollutants: 21
        2.5.2 Secondary pollutants: 21
    2.6   Ozone (O_3) 22
    2.8   Sulphur Dioxide 23
    2.7   Carbon Monoxide 24
    2.9   Oxides of Nitrogen 24
    2.10   Particulate Matter 25
    2.11   Hydrocarbons 25
    2.12   Sources and Causes of Air Pollutants 27
        2.12.1 Technology as a General Source 27
        2.12.2 Transportation and Air Pollution (The Mobile Sources) 27
        2.12.3 Solid Waste Disposal and Incineration 28
    2.13   Effects of Air Pollution 28
        2.13.1 Effects on Humans 29
        2.13.2 Effects on Animals 31
        2.13.3 Effects on Plants: 34
        2.13.4 Effects on Materials 36

Chapter 3: Introduction to Lahore City 37
   3.1   Introduction 37
   3.2   Climate 39
         3.2.1 Temperature 40
         3.2.2 Humidity 40
         3.2.3 Wind Direction.. 40
   3.3   Population Growth of Lahore... 40
   3.4   Existing Land Use 42

Chapter 4: Patterns of Air Pollution in Lahore City 45
   4.1   Introduction 45
         4.1.1 Wind speed... 46
         4.1.2 Wind direction 46
         4.1.3 Temperature 46
         4.1.4 Humidity 46
         4.1.5 Barometric pressure 46
         4.1.6 Sunshine 46
   4.2   Patterns of Ozone Concentration in 2000 48
   4.3   Patterns of Sulphur Dioxide in the city in 2000................................. 53
   4.4   Patterns of NO_x in the city in 2000 58
   4.5   Patterns of Carbon Monoxide in the City in 2000 63
   4.6. Patterns of Particulate Matter in the city in 2000.............................. 68

Chapter 5: People's Behaviour towards Air Pollution... 73
   5.1   Public survey 73

Chapter 6: Summary and Conclusion 95
   6.1   Summary.. 95
   6.2   Conclusion 96
   6.3   Recommendations: 97

Annex-1: Bibliography... 99

Annex-2: Bio of Authors 103

Annex-3: Questionnaire 105

Chapter 1:
Introduction

The atmosphere, which is surrounding the earth and providing us oxygen to breath, consists of nitrogen, oxygen, water vapour and inert gases. One of the major problems, that is arising concern among all quarters of the world, is the rapidly deteriorating atmosphere occurring due to the air pollution. Until recently, environmental pollution problems have been minor because of earth's own ability to absorb and purify minor quantities of pollution. However, due to the industrialization, the introduction of motorized vehicles, explosion of human population and the indiscriminate discharge of gases into the atmosphere by industries without considering the consequences have increased the magnitude and gravity of this problem.

Air pollutants are the substances in the atmosphere, which have harmful effects on human and biotic life. Organisms are able to deal with certain levels of pollutants without suffering ill effects. However, the pollutant level below which no ill effects are observed is called the "threshold level" (Nebel et. al., 1993).

The major sources of air pollution are power and heat generation, burning of solid wastes, industrial processes, and transportation. Bruce et al (2000) have pointed out three factors which determine the level of air pollution:-

1. Amount of the pollutant put into the air.
2. Amount of the space into which the pollutants are dispersed.
3. Mechanism that removes pollutants from the air.

1.1 Global Effects:

There are several major environmental problems such as smog, acid rain, the greenhouse effect and "holes" in the ozone layer which are the focus of the environment due to their harmful effects on the man and its environment.

Acid rain is one of the major environmental problems, which has always being the focus of the scientists. Chemical analysis of acid precipitation reveals the presence of sulphuric acid and nitric acid. In general about two-third of the acidity is due to sulphuric acid and one–third to nitric acid. Both sulphuric dioxide and nitrogen dioxide gradually react with water vapour and, through a number of steps, become acids. Precipitation becomes acidic, as it flushes these acids from the atmosphere. The pH of the precipitation depends on both the amount of acid and the amount of water. Fog and mist may be mostly acidic because the acid is disclosed in relatively little water.

Some scientists believe that inhalation of highly acidic fog particles and dry acid particles is a major source of breathing and respiratory problems (Dockery et. al., 1994; Thurston, 1991). In addition there is evidence that inhalation of such particles renders lung tissues more susceptible to the carcinogenic effects of to the pollutants.

Climate has always been affected by the human activities (Shafiq, Santos et. al., 2011). However with growth of population and their needs, such as land use and transportation the climatic conditions are being altered very rapidly in the present time. This alteration is being caused by the addition of carbon dioxide and certain other gases to the atmosphere, which causing the earth's climate to warm resulting in rising sea levels and insecure weather changes throughout the world.

This carbon dioxide effect is also known as 'the greenhouse effect', because it is analogues to solar heating, which occurs in the green house or in the car when it left parked in the sea.

Sunlight enters the glasses and absorbed. As the surface became warm, they radiate energy as infrared or heat radiation. The nature of the glass is that, it is highly transparent; it tends to block the infrared radiation. Therefore the energy that enters as light is trapped and causes the temperature to rise. The carbon dioxide acts as a heat trapper and initiates to increase the temperature level in the lower atmosphere.

On the global level, carbon dioxide in the atmosphere plays a role analogue to the glass. Light energy comes through the atmosphere. It is absorbed and converted into heat energy at the surface and exists as infrared radiation; the natural gases in the atmosphere do not. As carbon dioxide absorbs the infrared radiation, it becomes warms and thus in turn warms the rest of the atmosphere. Consequently, it follows that greater the amount of infrared that will be absorbed and the warmer will be the atmosphere. Climatologists are now in general agreement that if a doubling of carbon dioxide level to 600ppm is used as a reference point, the overall warming will be between 1.5 $^{\circ}$C and 4.5 $^{\circ}$C, warming in likely to be more pronounced in polar region as much as 10 $^{\circ}$C , and less pronounced in equatorial region 1 $^{\circ}$C – 2 $^{\circ}$C.

Depletion of ozone shield is a serious problem in the atmosphere. Radiation from the sun includes the ultraviolet light along with visible light. Ultraviolet is like light radiation but the wavelengths are slightly shorter than violet light, which are the shortest wavelengths seen by the human eye; while ultraviolet light is not visible, the rays are more energetic than those of visible light. On penetrating the atmosphere and being absorbed by biological tissues, they actually destroy protein and DNA molecule. If the full amount of ultraviolet radiation falling on the upper atmosphere came through the earth's surface, it is doubtful if any life could survive, plants and animals alike would simply be "cooked"; even the small amount that does come through is responsible for all the sunburns and some 200,000 cases of skin cancer per year in the USA (Miller, 2000)

We are spared more damaging effects from ultraviolet rays because most of it is absorbed and thus creamed out by a layer of ozone in the stratosphere. It is commonly referred to as ozone shield. There are various man-made pollutants that are causing it to breakdown. The most significant ones are free chlorine atoms, which are highly poisonous to plants and animals (Thomas, 1984).

1.2 Air Pollution in Pakistan

Air pollution is woven throughout the fabric of our life. A by-product of the manner in which we levied our cities, air pollution is the waste remaining from the ways we produced our goods, transport our goods and ourselves and generate energy to heat and light the places where we live. In terms of Pakistan 'Environmental act 1997', the air pollution is the release of any substance like soot, smoke, dust particles, odour, light, electromagnetic radiation, heat, fumes, communication exhaust, gases, noxious gases, noxious gases, hazardous substances and radioactive substances into the atmosphere to the extent, which have adverse environmental effects or on human health and safety and property, or on biodiversity (Barker & Tingey, 1992). Impurities in fuel, poor fuel air ratio, too high or too low combustion temperature cause the main pollution in the air. Industrial resources emit air pollutants through combustion of fuel, chemical processes, manufacturing, grinding, mixing, evaporation, and drying processes (Smith, 1993). The industrial units emit following air pollutants: carbon monoxide, carbon dioxide, nitrogen dioxide, and sulphur dioxide, organic vapour, and organic compounds etc. The mobile sources (Zhou & Levy, 2007) like automobiles, diesel-powered trains, trucks, and airplanes constitute more than 40% of the five major air pollutants.

Sector	1987/88		1997/98	
	CO_2	SO_2	CO_2	SO_2
Industry	26680	423	53429	982
Transport	10250	57	18987	105
Power	11216	95	53062	996
Domestic	24050	16	3998	40
Agriculture	4490	28	6368	40
Commercial	2587	13	4261	25

Table 1.1: Patterns of major pollutants in Pakistan Source: The Pakistan National Conservation Strategy (1999), Government of Pakistan

Table 1.1 shows that carbon dioxide generation from industry is very high and has risen to about 100% in the last decade. While power generation and transportation play the second and third roles respectively in producing high levels of air pollution.

City	PM_{10}	SO_2	CO	NO_2	HC
Karachi	6175	4224	3777380	44675	90584
Lahore	3213	2218	198180	23460	47570
Islamabad	1572	1075	96090	11375	2365
Faisalabad	1344	920	82130	9722	19714
Hyderabad	1148	785	70160	8305	16841
Multan	1094	784	66880	7917	15077
Peshawar	1028	703	62815	7436	7257
Quetta	495	338	30235	3579	-

Table 1.2: Annual vehicular emission in different cities [in tons] Source: EPA

A comparative study indicates that our vehicles emit 25 times more hydrocarbons and 3.6 times more nitrous oxide than those in the US. Nitrogen dioxide present in our environment is much more than UN WHO specified limit of 0.05 ppm.

1.3 Air Pollution in Lahore City

Lahore city is highly polluted city of Pakistan. The major causes of increase in air pollution in Lahore city are industries, more traffic, and high density of population. The major pollutants for this study are sulphur dioxide, oxides of nitrogen, ozone, and particulate matter.

Location	Ozone	SO$_2$	NO$_x$	CO	Dust
Yateem Khana Chowk	8	18	175	3	1123
Lohari gate	10	20	68	2	1180
Azadi chowk	12	25	125	1	850
Bank square	4	38	208	19	1050
Qartaba chowk	8	25	105	22	1030

Table 1.3: Air pollution in Lahore in 2000
Day time average Source: EPA

1.4 Aims and Objectives of the Study

The present research explores the extent of air pollutants in Lahore city. The main objectives are:

1. To observed the air pollution level in different parts of the city.
2. To see the spatial pattern of air pollution in Lahore city.
3. To see the differences in the air pollution among major parts of the city.

These objectives have been achieved through collecting primary and secondary data of about six major pollutants.

In order to evaluate the patterns of air pollution, the pollutants such as suspended particulate matter, sulphur dioxide, nitrogen oxides, ozone, and carbon monoxide have been examined.

1.5 Data Sources

The nature of study reflects that primary as well as secondary data have been used in this study.

For primary data collection it is very important to select a sampling site. So ten sampling sites have been chosen. A sampling point is selected if it meets the following requirements;

1. It is not directly exposed to varying amount of gas.
2. It produces a homogenous level of pollutants
3. Day conditions at the sampling point are consistent.
4. The point is located where the maximum people are expected to be affected by pollution

Some secondary data were also obtained.

The data for the carbon monoxide, nitrogen oxides, sulphur dioxide, and ozone in 1993, 1996 were available from Environment Protection Agency. Data of particulate matter in the above given years were available from University of the Engineering and Technology Lahore, while some data was also available from Space Science Department, University of the Punjab.

1.6 Methodology

As the aim of the research indicates that major emphases in the present study has been placed on the present day problem of air pollution and the changing patterns. The second chapter is the detailed description of the air pollutants, their sources causes and their effects. The third chapter is presenting the general picture of the physical geography of Lahore. To achieve the objectives of the research primary data were collected at sampling site. These are:

1. Yateem Khana Chowk
2. Yadgar Chowk
3. Charing Cross
4. Gulberg [Main Market]
5. Campus Bridge

6. Mall road [Regal Chowk]
7. Qartaba Chowk
8. Kot Lakhpat
9. Badami Bagh
10. Shadman chowk

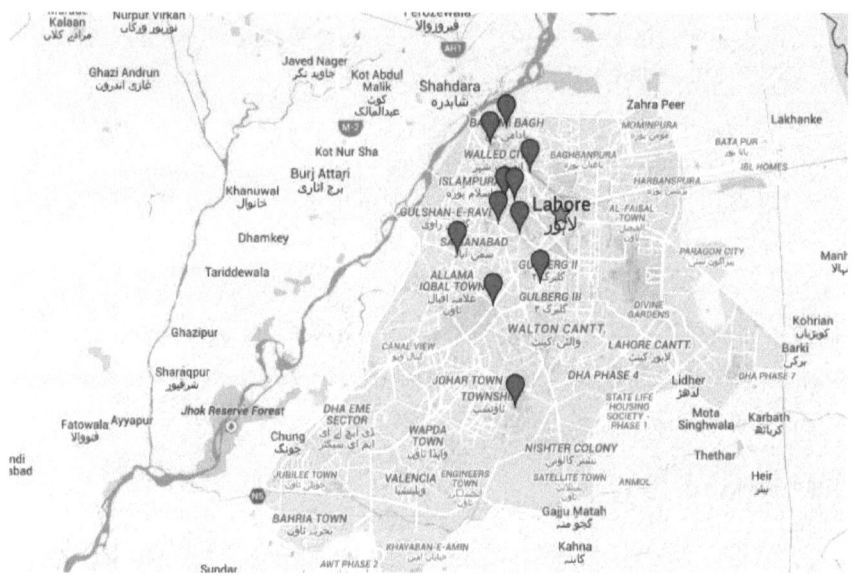

Figure 1.1: Map of Lahore city showing the urban extend [in dark shade] and sampling sites [in red balloons].

For this purpose, equipment known as the air pump was used with the help of Department of Space Science of University of the Punjab (Shafiq, 2001). The reagents for the detection of air pollutants were obtained and prepared from Institute of Chemistry, University of the Punjab.

The secondary data of these stations were easily available from the above given departments. Difficulties were faced in compiling the data because the data were in raw form and it took a lot of time to bring it in the desired form. Map of Lahore has been digitized to represent the primary and secondary data. Results have been presented in the form of the tables, graphs and maps, showing patterns of air pollution in Lahore city. The data have been shown in the form of temporal maps as shown in figures 4.4, 4.7, 4.10, 4.13, 4.16, and choropleth maps are in Figure 4.3, 4.5, 4.8, 4.11, 4.14 and in the form of Tables and Graphs.

The last chapter summarizes the whole research and gives the conclusion.

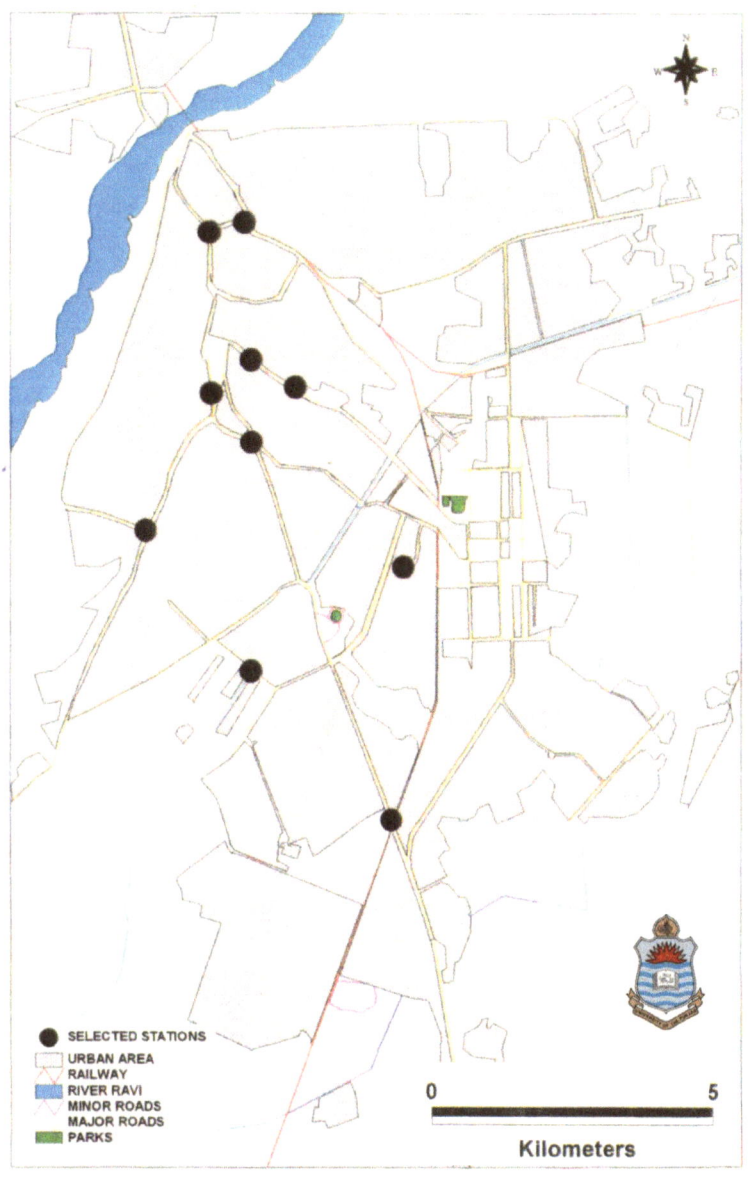

Figure 1.2: Map of Lahore city showing sampling sites loaded into a GIS

Chapter 2:
Air Pollutants, Source and their Effects

2.1 Introduction

In simple words any addition to air, water, soil, and food that threatens the health or other living organisms can be called pollution. Most pollutants are solid, liquid or in gaseous form by products or wastes of industrial material which are produced, when a resource is extracted, processed, made into products, or used (Nebel et. al., 1993; Smith, 1993). Pollution can take the form of unwanted energy emissions, such as successive heat, noise or radiation. Most pollution from human activities occurs in or near urban and industrial areas, where pollutants are concentrated.

Environmental pollution may also be defined as "the unfavourable alteration of surroundings, wholly or largely as a by-product of man's actions, through directed or indirect effects of changes in energy patterns, radiation levels, chemical and physical constitutions and abundance of organisms" (Nebel et. al., 1993). Such changes may affect man directly or through his supplies of water and of agricultural and other biological products, his physical objects or possessions, or his opportunities for recreation and appreciation of nature.

Whenever society uses energy and material there are both desirable and undesirable products. The desirable products are the discarded materials and energy given up in the course of performing work. Most environmental pollution comes from these by-products. In older times it was heat, smoke, and related wastes from one's home, which have been contributed to air pollution. But with the development of technology, the contribution to air pollution has changed dramatically and the nature of the pollution has changed accordingly.

2.2 Types of Pollution

Usually pollution is classified according to the environment in which it occurs or according to the types of pollutants [lead, mercury, carbon dioxide, solid wastes, noise, biocide, heat, etc.] by which pollution has been caused. Sometimes, pollution is made to classify into two broad categories

» Artificial pollution, which originates due to the activities of man.
» Natural pollution, which originates from natural processes.

However, keeping in mind the major types of pollution can be categorized as:
1) Air pollution
2) Water pollution
3) Solid waste pollution
4) Land pollution
5) Marine pollution
6) Noise pollution
7) Radiation pollution
8) Thermal pollution

The focus of this research is on air pollution.

2.3 Air Pollution

Air pollution is defined as "an atmospheric condition in which certain substances are present in such concentrations that they can produce undesirable effects on man and his environment" (Kupchella, 1993).

These substances are gases, particulate, radioactive materials, etc. This means gases or aerosol material in the air, which are not considered to be a normal constituent or excess of normal constituent. In simple words "Air pollution is basically the presence of foreign substances in air" (Kupchella, 1993). Critics have pointed this out that the rapid industrialization is the major cause of air pollution. The air pollution can be understood as the "price of industrialization" and air pollution caused by automobiles as the "disease of wealth".

Air pollution means the presence on the outdoor atmosphere of one or more contaminants such as dust, fumes of gases, mist, odour, smoke, or vapour in quantities, which are injurious to human, plants, animal life, property or which quantities, which unreasonably interfere with the comfortable enjoyment of life and property. There is no doubt that air pollution existed in ancient times, but it was much less widespread than today. Today air pollution level has increased enormously at regional and global level.

Pollutants	Composition	Characteristics
Sulphur dioxide	SO_2	Colourless, heavy, water soluble, gas with a pungent smell, irritating odour
Particulate matter	Variable	Solid particles or liquid droplets including fumes, smoke, dust, and aerosols
Nitrogen dioxide	NO_2	Reddish brown gas, somewhat water soluble
Hydrocarbons	Variable	Many and varied compounds of hydrogen and carbon
Carbon monoxide	CO	Colourless, odourless, slightly water soluble
Ozone	O_3	Pale blue gas, fairly water soluble, unstable, sweetish odour
Hydrogen sulphide	H_2S	Colourless gas with smell of rotten eggs
Fluoride	Variable	Pungent, colourless, water soluble gas
Nitric oxide	NO	Colourless gas, slightly water soluble
Lead	Pb	Metallic, can exist in variety of ways
Mercury	Hg	Metallic, can exist in variety of chemical compounds

Table 2.1: Molecular Composition and characteristics of major pollutants

2.4 Air Pollutants

In general term these are the additions that contribute to the air pollution. Air pollutants are the physical, chemical, biological and radioactive materials, which make the air undesirable for breathing. These are of worldwide range and variety. However, they can be categorized in two groups.

2.5 Categories of Air Pollutants

2.5.1 Primary pollutants:

These pollutants are directly emitted into the atmosphere from the terrestrial sources. These include the gases such as carbon dioxide, carbon monoxide, Sulphur dioxide, ammonia, hydrogen sulphide, nitric oxide, hydrogen fluoride, particulate like dust, smoke, ash and fumes, radioactive substances and hydrocarbons. Primary pollutants include pollen particles, salt-water spray, wind-blown dust and fine debris from volcanic eruptions. Most man made pollution involves the products of combustion smoke, carbon monoxide and lead, and oxide of nitrogen and sulphur dioxide-through other industrial processes, crop spraying and atmospheric nuclear explosions also contribute.

2.5.2 Secondary pollutants:

These are formed from chemicals and photochemical interaction between primary pollutants and other atmospheric constituents. sulphur trioxide, nitrogen dioxide, sulphur nitrates, aldehydes and ozone are cited as examples.

These generations depend on atmospheric, topographic and meteorological conditions and the nature and concentration of the primary pollutants and the atmospheric constituents.

Most air pollution arises in the urban environment, with a large portion of that coming from automobiles. Six major types of substances, known as primary pollutants, account for more than 90% of the nation-wide air pollution. These six pollutants are (Sami, Ali et. al., 2004; Stern, 1984):

1) Carbon monoxide [CO]
2) Nitrogen oxides [NO$_x$]
3) Sulphur dioxide [SO$_2$]
4) Particulate matter [PM]
5) Hydrocarbons [HC]
6) Ozone [O$_3$]

2.6 Ozone (O$_3$)

Ozone is a gas of blue colour. The combination of hydrocarbons and oxides of nitrogen in automobile exhaust and sunshine leads to the formation of ozone and other photochemical oxidants. A host of products are generated in the atmosphere through the interaction of nitric oxide, sunlight and hydrocarbons. These include ozone, peroxyacetyl nitrate, and acrolein. Oxygen molecules can absorb ultraviolet radiation directly, causing them to split into two atoms of oxygen that eventually go on to form ozone, this occurs to significant extent only high in the atmosphere.

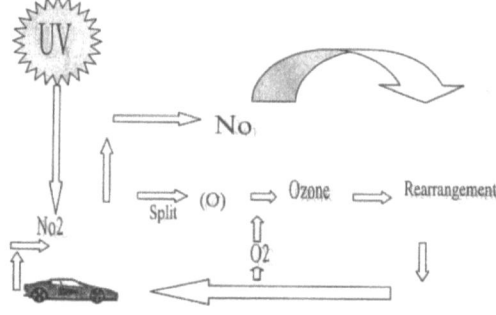

Table 2.1: Formation of Ozone

Another mechanism is necessary to create the highly reactive atomic form of oxygen and then ozone in the air that people actually breathe. Nitrogen dioxide is a very efficient absorber of the ultraviolet light that does not reach the earth's surface. As nitrogen dioxide absorb such radiation, it is broken down into NO and O. This O combines with the O_2 to form O_3.

In cities, 95-98% of the carbon monoxide in the air is from human sources and the levels of carbon monoxide are many times higher than the average level in the natural world. Carbon monoxide averages about 0.1 ug/m^3 of air, concentrations may reach levels of 80-150mg/m^3 in heavy traffic. Prolonged exposure to concentrations as low as 50 ppm can impair judgment and reflexes (Kupchella, 1989).

2.8 Sulphur Dioxide

Sulphur dioxide is a gas poisonous to both plants and animals. Coal burning, electrical power plants are blamed for producing most of the sulphur dioxide problem. On the average, 70% of the sulphur dioxide in the air over cities comes from the utilities. Fuels vary greatly in their sulphur content. High sulphur coal from certain locales might have as much as 5% sulphur. Natural gas contain only a trace amounts of sulphur; this is why, when controls were first placed on sulphur dioxide emissions, many plants, factories, and power generating stations switched from coal to natural gas. Sulphur dioxide gas is itself a poison, but it can react with ozone, hydrogen peroxide, water vapour, and other substances in the atmosphere to form sulphuric acid. It is one of the strongest acids known; it is able to corrode limestone, metal and clothing, and it have a devastating effect on delicate respiratory tissue. Its threshold level is 0.5 to 1.0 ppm and level at which even normal people experience bronchial spasms is 1.0 ppm and EPA's 24-hour standard is 0.14 (Kupchella, 1989).

According to Environment Protection Agency, the probable median health effects, on which the ozone standard of 0.12 ppm was based, are that 0.15 ppm produces chest discomfort, irritation of respiratory tract, and reduction of pulmonary functions; at 0.17 ppm, asthma, emphysema, and chronic bronchitis are aggravated;

and at 0.18 ppm there is reduced resistance to bacterial infection. The allowable concentration of ozone in industry is 0.05 ppm. However, in some industries, concentration of ozone can occasionally reach nearly 1 ppm; this level is also sometimes reached in acute air pollution episode and causes major problem as given in the Table 2.2.

2.7 Carbon Monoxide

It is colourless, odourless, and tasteless gas slightly lighter than air. Everyone is aware that an idling automobile produce considerable amount of this pollutant. Human beings contribute about 10% of carbon monoxide load dumped into the atmosphere and nearly all of this comes from incomplete combustion, largely in the automobile of fuel. Naturally produced carbon monoxide comes mixed with methane and other substances in marsh gases and other gases emitted from decaying material. Carbon monoxide also escapes from forests and grass fires and volcanoes. Some formed through chemical reactions in the upper atmosphere. Human generated carbon monoxide is a problem because most of what we generate is dumped into the areas in which we also live and breathe.

2.9 Oxides of Nitrogen

When air is fed into a combustion mixture, particularly when the combustion is occurring, at a temperature above 2000 °F, the oxygen and the nitrogen present in the air as O_2 and N_2 combines to form nitric oxide. Nitric oxide is not thought to be very harmful and does not do much damage because it cannot readily dissolve in water in tissues. However, through the action of sunlight, nitric oxide can combine with oxygen to form nitric oxide. Nitrogen dioxide is a radish-brown toxic gas that has considerable environmental impact Nitrogen dioxide is similar to sulphur dioxide in many reaction with substances in the atmosphere, nitrogen dioxide is converted into inorganic nitrates, peroxyacetyl nitrate, and nitric acid. Natural agents like soil bacteria produce far greater amounts of the oxides of nitrogen than humanity does by its fires. Nitrogen dioxide is produced in the combustion of coal, oil, natural gas, and motor vehicle fuel and whenever temperatures are high enough to cause atmospheric nitrogen and oxygen to combine. EPA's quality standard for oxides of nitrogen is 0.05 ppm (Kupchella, 1989).

2.10 Particulate Matter

Small solid particles and liquid droplets, collectively called as particulate, are present in the air in great numbers, and at times they constitute a serious problem. However, particulate pollution has multiple components, including sulphate salts, sulphuric acid droplets, salts of metals, dust from finely divided particles of carbon or silica, liquid sprays and mist. The size of particulate matter is an important characteristic. Individual particulate are measured in micrometres. Particulates range in size from 0.005 um to about 100 um.

2.11 Hydrocarbons

The number of hydrocarbons involved in the air pollution is fairly large. Natural sources contribute huge quantities of these substances. Plants, particularly trees emit hydrocarbons. Most of the plants, which produce terpenes, belong to the coniferous family and the mytaceace family.

Automobiles produce most of the pollutant hydrocarbons added to the atmosphere through the activities of man. Some important air pollutants, their anthropogenic sources and their health effects have been described in table 2.2.

Air pollutants	Major anthropogenic sources	Effects on human beings
Particulate	Industrial combustion of fuels	Toxic effects through several mechanisms including interference with respiratory tract
Sulphur dioxide	Combustion of fuels in industrial processes	Irritation of respiratory system diminished lung function, aggravates asthma
Carbon monoxide	Transportation and agricultural burning	Toxic, easily enter blood with increasing concentrations, headache
Nitrogen dioxide	Transportation	Toxic respiratory tract problems with increasing concentration, nasal irritation breathing discomfort, death of animals
Ozone	Secondary pollutants derived from reactions with sunlight and oxygen	Toxic with increasing concentration, nose and throat irritation fatigue, lack of coordination.
Hydrocarbons	Automobiles, Industrial processes, Evaporation of organic solvents, Agricultural burning	Irritate mucous membrane

Table 2.2: Summary of sources and health effects of major air pollutants

2.12 Sources and Causes of Air Pollutants

Nature puts greater quantities of bad things into the air that human do. Sulphur dioxides, oxides of nitrogen, carbon dioxide, methane, hydrocarbons, and particulate emanate from volcanoes, swamps, forests, natural fires, and the action of wind on soil not covered by vegetation.

Volcanoes contribute great amounts of dust and particulate matter as well as noxious gases. Organic decay in tropical forests and other places puts a variety of gaseous compounds like methane into the atmosphere, and forests contribute various organic compounds including ketones, aldehydes and other complex hydrocarbons, which can participate in the generation of ozone.

2.12.1 Technology as a General Source

Modern technology is more costly, more energy consuming and tend to pollute more that the technologies it has replaced over the years. Beyond the impression that we do lead somewhat better and happier lives today, close inspection seems to reveal that we have to pay more, in terms of energy and deterioration of the environment, for equivalent units of economic goods produced than we did in decades past. Electrical power generation is responsible for the significant fraction of air pollution problems.

2.12.2 Transportation and Air Pollution (The Mobile Sources)

Human generated air pollution sources can be divided into mobile and stationary sources. Stationary sources include factories, incinerators, and other kind of non-mobile sources.

Mobile sources tend to be much smaller, much more plentiful, and much more widely dispersed than stationary sources and are difficult to monitor.

Automobiles and trucks are the main mobile sources problem with respect to air pollution because:

- » To carry the same load, trucks and cars emit about six times as much pollution as rail roads.
- » Automobiles produce many more times the amount of pollution per gallon of fuel consumed than diesel-powered trucks or trains or jet aircrafts. Overall, highway vehicles generate roughly 10 times more air pollutants than other mobile sources.

The burning of coal and other fuels are the principle cause of air pollution coming from stationary sources in many countries.

2.12.3 Solid Waste Disposal and Incineration

Package of products ranging from food to toys contributes significantly to the air pollution problem. First, pollution is generated in making all packaging and then after a package is removed, the package must be disposed-off often by incineration.

2.13 Effects of Air Pollution

Substantial evidence has accumulated that air pollution affects the health of human beings and animals, damages vegetation, soil and deteriorates materials, effects climates, reduces visibility and solar radiation, contribute to safety hazards and generally interferes with the enjoyment of life and property. Although some of these effects are specific and measurable, such as damages to vegetation and materials, and reduced visibility, many are difficult to measure such as health effects on human beings and animals and interfere with comfortable living.

2.13.1 Effects on Humans

Air pollution is one of the greatest environmental evils. The air we breathe has not only life supporting properties but also life damaging properties. Under ideal conditions the air we inhale has a quantitative and qualitative balance that maintains the wellbeing of man. But when the balance among the air components is distributed, or in other words, if it is polluted, it may effect human health. An average man breathes 22,000 times a day and takes in 16 kg of air each day. It far exceeds the consumption of food and water. It has been estimated that a man can live for five weeks without food and five days without water, but only for five minutes without air. Therefore the prime factors affecting human health are:

1. Nature of the pollutants
2. Concentration of the pollutants
3. Duration of exposure
4. State of health of the receptor
5. Age group of the receptor

The effects of the air pollution on human health generally occur as a result of contact between the pollutants and the body. Normally, bodily contact occurs at the surface of the skin and exposed membranes. Surfaces are of utmost importance because of their high absorptive capacity compared to that of the skin. Air borne gases, vapours, fumes, mist and dust may cause irritation of the membranes of the eyes, nose, throat, larynx, trachio-branchial tree and lungs. Following are the diseases have been noted as a result of air pollution:

1. Eye irritation
2. Nose and throat irritation
3. Irritation of the respiratory tract
4. A variety of particulate particularly pollens, initial asthmatic attacks

5. Carbon monoxide combines with the haemoglobin in the blood and consequently increases stress on those suffering from cardiovascular and pulmonary diseases
6. Hydrogen fluoride causes diseases of the bone, and mottling of teeth
7. Carcinogenic agents cause cancer
8. Certain heavy metals like lead may enter the body through the lungs and cause poisoning

Air pollution can originate of daily mortalities due to exposure to lower levels of airborne particles. The pollutants give rise to health problems associated with lung functions and more people gets hospitalized (Schwartz, 1994; Anderson, 1999).

The biological effects of radiation may be somatic or genetic damage. In somatic damage, the exposed individual is affected, in genetic damage the future generations becomes the victims. Radioactive fallout from testing of nuclear weapons causes:

1. Cancer
2. Shortening of life span
3. Genetic effects or mutation

One significant point we have to note about the effect of radioactive fallout is that it causes long-range effects affecting the future of man and hence the future of our civilization. Certain diseases typically associated with occupational air pollution show the same synergistic relationship with smoking. For example, Black lung disease is seen exclusively among coal minors, who are also smokers. Lung disease among those exposed to asbestos predominates in smokers.

Emphysema is a condition in which the alveoli in the lung become uneven and over distended due to destruction of the alveolar walls. This disease is accompanied by shortness of breath, particularly following exercise. The destruction of alveoli is progressive, resulting in an increased blood flow necessary to accomplish oxygen transfer and to a decreased ability to eliminate foreign bodies that reach the alveolar region. Emphysema has no known cure and one of the fastest growing causes of death in the United States.

2.13.2 Effects on Animals

Interest in effects on air pollution on animals has generally developed as a corollary to concern about its influence on human health. Most of the information concerning the natural exposure of animals to air pollution is contained in the reports of some major air pollution disasters e.g. Donora, London and Poza Rica. Recently considerable information has been reported from medical research laboratories which describe the results of experimental exposure of small animals to various air pollutants, rats guinea pigs and monkeys.

Concentration [ppm]	Species	Effects
0.25	Rabbit	Irreversible lung collagen Change
0.5	Rat	Lung mass cell degranulation
0.5	Mouse	Alveolar wall thickening
0.8	Mouse	Pulmonary alveolar edema
0.8	Rat	Terminal bronchiolar hypertrophy
1.0	Rat, Dog, Rabit	Focal interstitial pneumonia
1.5	Mouse	Desquamaire bronchitis
2.0	Monkey	Bronchial - Ronchiolar hyperplasia and metaplasia

Table 2.3: Histopathological effects of inhaled NO_2 in experimental animals. Source: (Kupchella, 1989)

The process by which farm animals get poisoned is entirely different from that by which human beings exposed to polluted atmosphere are poisoned. In case of farm animals it is a two steps process:

1. Accumulation of air-borne contaminants in the vegetation and forge.
2. Subsequent poisoning of the animals when they eat the contaminated vegetation.

The three pollutants responsible for most livestock damage are fluorine, arsenic, and lead. These pollutants originate from industrial sources or from dusting and spraying.

Of all farm animals, cattle and sheep are the most susceptible to fluorine toxicities. Horses appear to be quit resistant to fluorine poisoning. Poultry are probably the most resistant to fluorine, of all farm animals. Lack of appetite, rapid loss in weight, decline in health and vigour, lameness, Periodic diarrhoea, muscular weakness and death, and characterize the acute form of fluorine poisoning. It may also result in considerable increase of borne fluorine. But acute poisoning due to fluorides is unlikely in a majority of the cases. Fluorine is a cumulative poison under conditions of continuous exposure to sub-acute doses. Fluorine is also a protoplasmic poison. It has a marked affinity for calcium and interferences with normal calcification. Animals have been reported to be more resistant than humans to dental mottling. Cattle and sheep are the most frequently affected. Hence tooth symptoms are a sensitive and unique criterion of chronic fluorosis. Excessive wearing of incisor and molar teeth may occur at high levels of fluorine intake.

Arsenic air pollution has caused deleterious effects on animals near smelters processing arsenic ores.

Diseases	Host	Remarks
Bovine Tuberculosis	Cattle, Swine, sheep, dog, cats	Contagious to man. Control by slaughter
Glanders	Horse, Mules	TB-like nodules and ulcers in respiratory tract, internal organs and skin. High death rate.
Aspergillosis	Birds, Ducks	Inhaled from grains and straw contaminated with mold.
Cryotococcosis	Horse	Causes granuloma in horses
Coccidiomycosis, valley fever, desert	Cattle, Horse, Swine, Dogs	Varies in severity symptoms of common cold
Histoplasmosis	Domestic animals	Causes lung lesions from inhalation of spores

Table 2.4: Airborne bacterial and fungal diseases of animals.

2.13.3 Effects on Plants:

Air pollution has long been known to have an adverse effect on plants. It has shown to reduce the yield of major crops in the EU and US but little is known about it in South Asia particularly in Pakistan (Khan & Siddiqui, 1982; Maggs et. al., 1995; Wahid et. al., 1995a; Wahid et. al., 1995b). *Triticum aestivum* L. (Wheat) and *Oryza sativa* L. (Rice) are among the major crops in the Pakistan whose economy is based on agriculture. Wahid et. al., (1995a) and Wahid et. al., (1995b) have demonstrated the reduction in the yield of wheat by 46.7% and rice by 42% in a high pollutant setting at Lahore.

At first, it was only sulphur dioxide that was considered a dangerous pollutant. Now with the advent of various pesticides and new industrial processes, the range of harmful pollutants has multiplied tremendously. Sometimes, vegetation over 150 km away from the source of the pollutant has been found to be affected. Industrial pollution, particularly from smelters, has caused complete destruction of vegetation in some cases.

The primary factor, which controls gas absorption by the leaves, is the degree of opening of the stomata. When the stomata are wide open, absorption is maximum and vice versa consequently, the same conditions that enhance the absorption of the gas predispose the plant to injury.

Most plants close their stomata at night and are therefore much more resistant at night than in the daytime. But some plants like the potato, which do not close their stomata at night, are as sensitive in the dark as in the night. A number of pollutants given below affecting the plants in one or another way:

1. Sulphur dioxide
2. Fluoride compounds
3. Ozone
4. Chlorine
5. Hydrogen chloride

6. Nitrogen oxides
7. Ammonia
8. Hydrogen sulphide
9. Hydrogen cyanide
10. Mercury
11. Ethylene
12. Peroxy acetyl nitrate
13. Herbicides [Sprays of weed killers]
14. Smog

The above pollutants interfere with plant growth and the phenomenon of photosynthesis. Smog, dust etc. reduce the amount of light reaching the leaf and also by clogging the stomata may reduce the carbon dioxide intake to some extent and thus interfere with photosynthesis.

Damage to leaves takes several forms:

- Necrosis: Necrosis is the killing or collapse of tissue.

- Cholorosis: Cholorosis is the loss or reduction of the green plant pigment, chlorophyll. The loss of chlorophyll usually results in a pale green or yellow pattern

- Abscission: Leaf abscission is dropping of leaf

- Epinasty: Leaf epinasty is a downward curvature of the leaf due to higher rate of growth on the upper surface.

The effects of various pollutants like sulphur dioxide, ozone, fluorides, etc. on plants is shown in the following Table.

Pollutants	Dons	Effects
Sulphur dioxide	Mild Severe	Interveinal chlorotic bleaching of leaves Necrosisin interveinal areas and skeletonized leaves
Ozone	Mild	Fleck on upper surfaces premature aging and suppressed growth
	Severe	Collapse of leaves, necrosis and bleaching
Fluorides	Cumulative effect	Necrosis at leaf tip
Nitrogen dioxide	Mild	Suppressed growth, leaf bleaching
Ethylene	Mild	Epinasty, leaf abscission
PAN	Mild	Bronzing of lower leaf surface, suppressed growth. Young leaves are more susceptible

Table 2.5: Effect of Pollutants on Plants.

2.13.4 Effects on Materials

Air pollution can affect materials by soiling or chemical deterioration, high smoke or particulate levels are associated with soiling clothing and structures and acid or alkaline particles, especially those containing sulphur, corrode materials such as paints, electrical contacts, and textiles. Ozone is particularly effective in deteriorating rubber. Heating systems and power generating stations have been identified as the source of the Sulphur dioxide that is believed to affect the material.

Chapter 3:
Introduction to Lahore City

3.1 Introduction

Lahore is a typical inland city of Pakistan. It is situated in the northward loop of the river Ravi. It lies between 31º 15' and 31º 42' North latitude and 74º 39' East longitudes. The Ravi passes around the North of the Lahore. In the North and the East, Lahore is bounded by Sheikhupura district, in the South by Kasur and in the East by India.

The total area of Lahore is 1,772 km (Mazhar & Jamal, 2009; Shirazi, 2012). Today Lahore is serving as a provincial capital of Punjab. Lahore is the hub of culture. It is said to have been founded by Loh, the son of Rama Chandra in 8th century AD. In 1021-23 AD the city was occupied by Mohammad Ghaznavi but Lahore, as we know it today, reached the peak of glory during the reign of Mughal rulers, especially during the time of Akbar The Great, who made it his capital.

Lahore occupies a fertile alluvial plain formed by the deposits of the river Ravi and Sutlej and their tributaries. The river flows along the north-western boundary of the city, and its seasonal floods have inhabited the city's growth towards the North and Northwest.

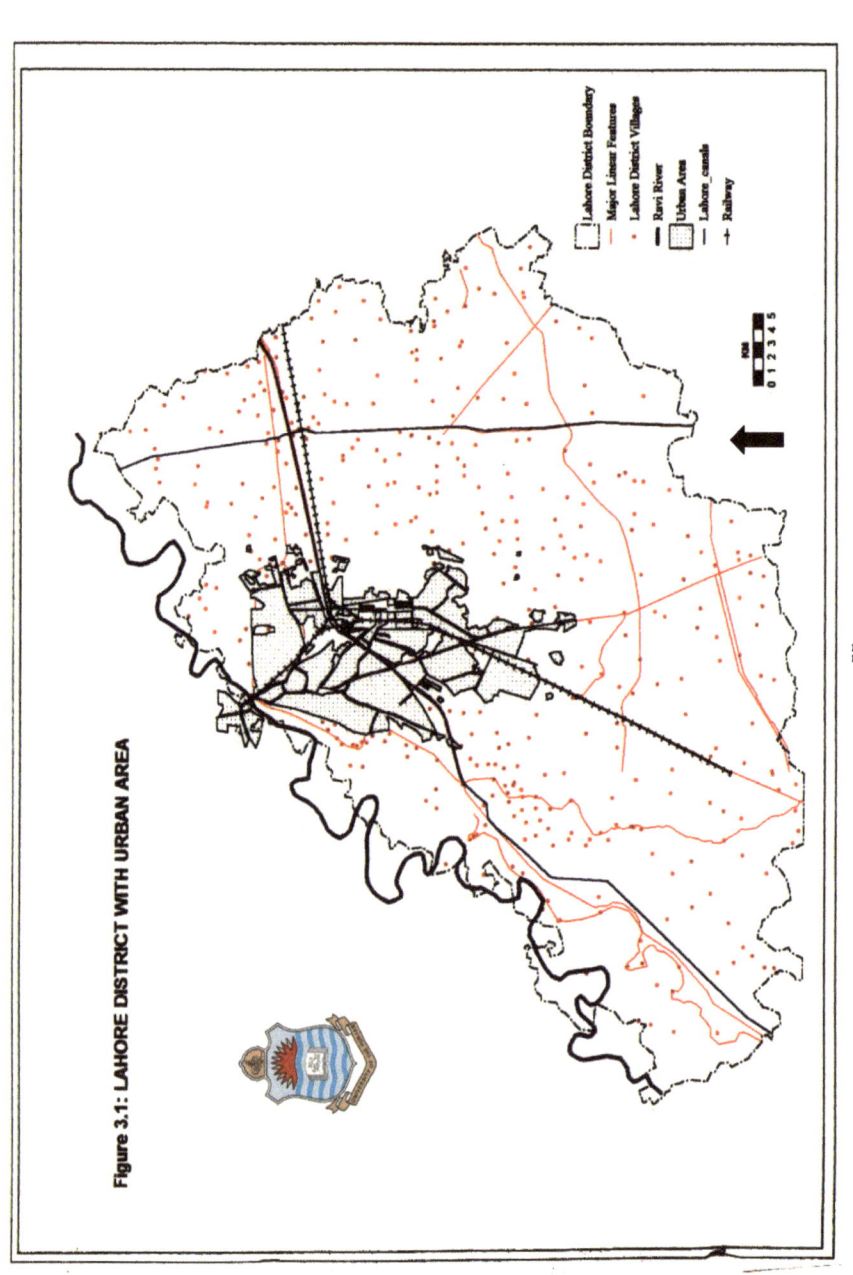

Figure 3.1: LAHORE DISTRICT WITH URBAN AREA

3.2 Climate

Climatically Lahore experiences three seasons, cold, hot, and rainy. The cold season extends from mid-November to the mid-February. In this season the cold sunny days are alternated with cloudy, rainy days. The rainfall is sight and takes place in the form of drizzle showers and may extend over a week. Rainy days are usually followed by clear days with frost at night in the countryside. Early morning and evening mist is also common.

The winter rain usually stops by the end of February but sometimes continue through March. The monsoon burst over Lahore sometimes in the mid of the June or the last week of the June. It usually results in the heavy downpour and the showers continue with the short intervals till September. The monsoons are at their wettest in July. In the rainy season, it is pleasant, but during the day following rains, the still air, high temperature, and high humidity create a condition of extreme discomfort. On the average there is no month without rainfall in Lahore. The driest months are October and November, which together receive on an average 0.34 inches of the rainfall. From December to March is the season of winter rains, which the entire four months period receive 3.72 inches of rainfall an amount although meagre, but of great importance for its value to agriculture and the weather. April and May receive rainfall either through western disturbances or locally generated thunderstorms. May to October account for only about 25% of the annual rainfall. The remaining 75% takes place in the monsoon season.

The monsoons are at their peak during July and August. On the average July is the month of the maximum and November is of minimum rainfall. The most intensive recorded cloudburst in Lahore occurred on September 24, 1954 which about 9 inches rainfall in 24 hours. Such cloudbursts seriously affect the daily life of the city by flooding the low-lying areas due to inadequate drainage system. The heaviest rainfall on Lahore occurred in the 1882 when recorded 37.43 inches of rainfall was recorded. The driest year was 1899 when only 6.21 inches of rainfall.

3.2.1 Temperature

Temperatures remain comfortable in this zone. The average of minimum temperature of January is 66 °F. On very rare occasions the night temperature falls below freezing point. The lowest temperature ever recorded is 40 °F May and June constitute the hot season of which June is the hottest month. The day temperature rises to scorching level, and due to hot breeze, is instantly uncomfortable. The average minimum temperature in June is 80 °F, though the highest of the season may touch 116 °F mark. In July, during the day following rains, high temperature is experienced. January is the coldest month.

3.2.2 Humidity

Humidity is very high in monsoon season, especially on the day following rain; high humidity is experienced in Lahore, which create a condition of extreme discomfort.

3.2.3 Wind Direction

The wind direction in Lahore changes from day to day and season to season. However, on the average there is a change of general direction from winter to summer when there is a complete reversal of the prevailing winds from November to May, the predominant wind direction is from West to East. During September and October, wind is changing its direction and is in the state of flux. By and large, it may be said that the predominant wind direction in Lahore is north-west.

3.3 Population Growth of Lahore

The first few years after the independence, in spite of large number of migrants from India and a significant increase in the population, did not see much of the development in Lahore. The economic development of the recent years has brought about acceleration in the rate of spatial expansion of the city. Lahore has grown almost 14 times in size since the beginning of the 20th century.

Population of the Lahore city has been increasing rapidly from 1951-1998 (Mazhar & Jamal, 2009). Estimates of population for the year 1991 and 1998 also show the same trend. Lahore is densely populated city. The growing population is exerting more and more pressure on the transportation of the city and creating severe problem of air pollution.

Census year	Population (millions)
1951	1.135
1961	1.626
1972	2.588
1981	3.545
1998	6.213

Table 3.1: Population of Lahore City

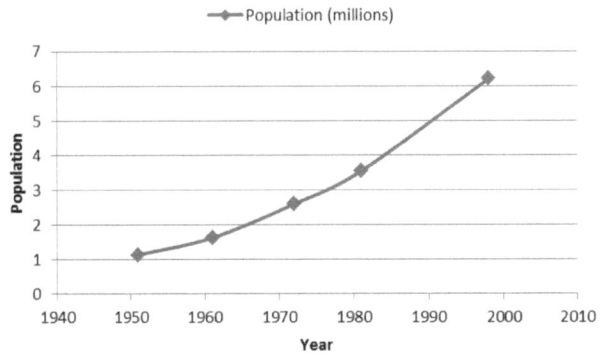

Figure 3.2: Population growth of Lahore City

3.4 Existing Land Use

People residing in the cities put land to different uses. The requirement of our study is to show that how the patterns of air pollution vary in different zones of the city. The commercial centre of Lahore is located along the Mall road. New commercial areas are mixed with the planned residential areas in the Southern part of the UBD canal. Mostly large factories are located along trunk roads in the suburbs and concentrated along Sheikhupura road and G.T. road.

There are a number of steel bar plants located North of the Badami Bagh bus terminal and car body manufacturing plants along band road and Multan road. Traditional houses are located in and around the walled city. A mixed use industrial and residential area is situated along the western portion of the Band road. Well-planed housing areas are main Gulberg, Model Town, and Defense. Low-density residential areas are located in the cantonment. Parks, for example; Minar-e-Pakistan and Bagh-e-Jinnah are scattered all over the urban area.

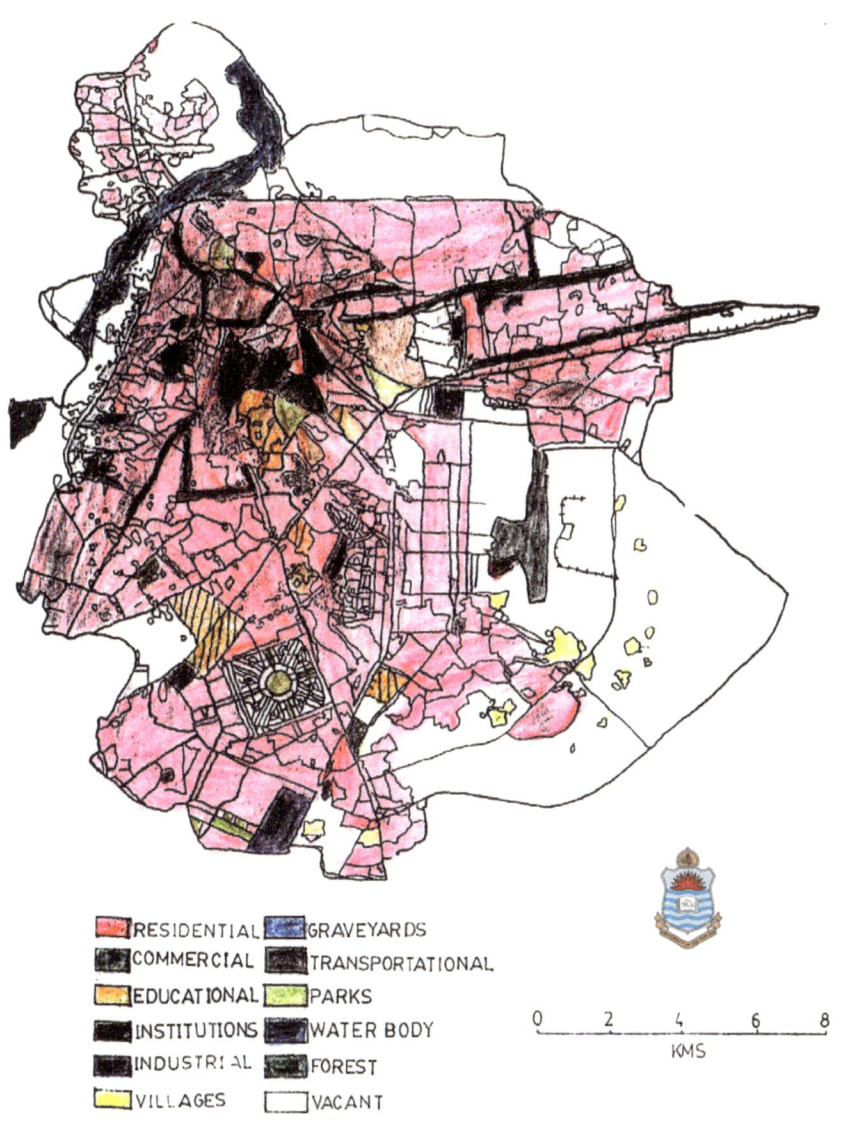

Figure 3.3: Land Use of Lahore City

Left blank for pagination

Chapter 4:
Patterns of Air Pollution in Lahore City

4.1 Introduction

Air pollution is the contamination of the atmosphere by harmful vapours, aerosols and dust particles, resulting principally from the activities of man but to a lesser extent from natural processes.

Primary pollutants include pollen particles, salt water spray, wind-blown dust and fine debris from volcanic eruption. Most man made pollution involves the products of combustion smoke, carbon monoxide and lead, and oxides of nitrogen and Sulphur dioxide-through other industrial processes, crop spraying and atmospheric nuclear explosions also contribute. Most air pollution arises in the urban environment, with a large portion of that coming from automobiles.

Lahore is a polluted city with a very high population density. The value of concentration of pollutants differs from place to place in Lahore City. Badami Bagh, Bhatti Chowk, Yadgar Chowk, Chowk Yateem Khana are the densely populated areas of Lahore.

All the readings have been shown on the map according to the data available. It is important to note that concentration of any gas is effected by the following factors given below:

4.1.1 Wind speed

It is very important in determining the level of any pollutant because if wind speed will be high at the sampling site, all the pollutant may get transported somewhere else from the sampling site and it can disturb the accuracy.

4.1.2 Wind direction

It also affects the accuracy of the reading because if the reading is taken in the opposite side of the wind direction the level of pollutant may deviate from the actual level.

4.1.3 Temperature

Temperature is the main factor in determining the level of air pollutants. If the temperature will be very high, it can speed up the reaction of pollutants.

4.1.4 Humidity

High humidity can disturb the concentration of any pollutant.

4.1.5 Barometric pressure

Barometric pressure is also noted because temperature and pressure are inversely proportional to each other. So any change in the pressure may change the reading.

4.1.6 Sunshine

Usually readings are taken during the daytime in the presence of sunshine because in cloudiness humidity increases and gas particles tend to cluster and disturb the accuracy.

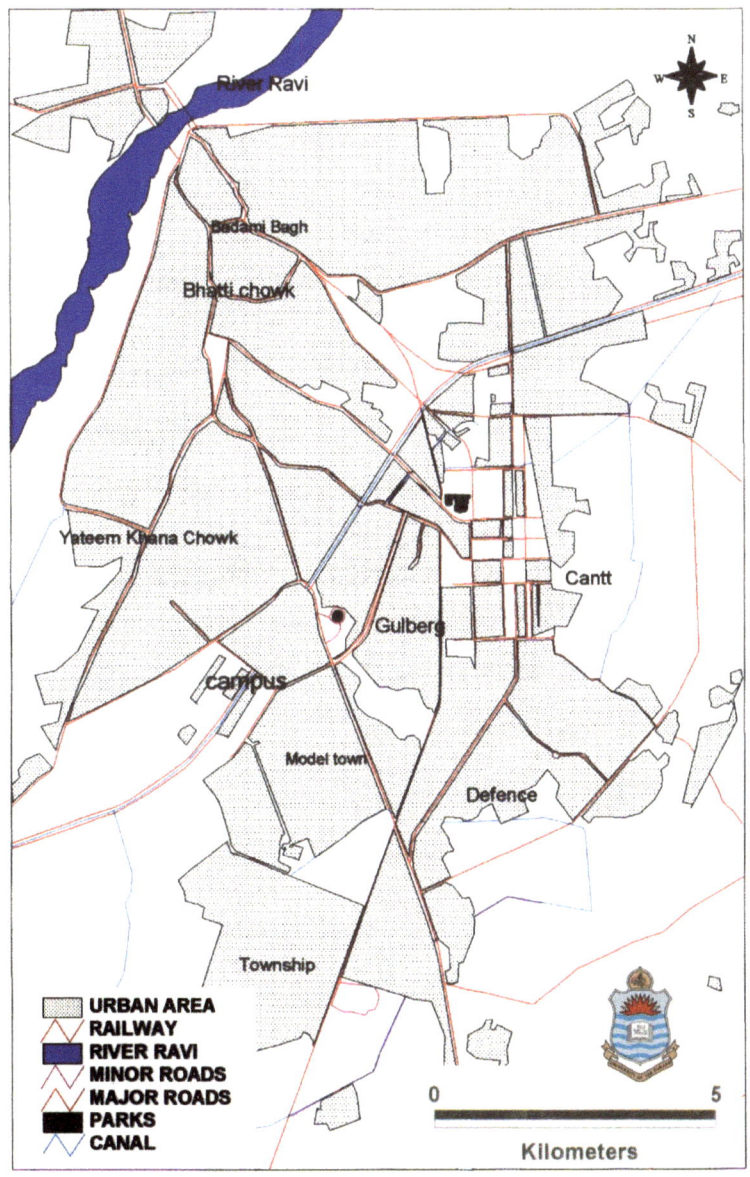

Figure 4.1: The Lahore City

4.2 Patterns of Ozone Concentration in 2000

Fig.4.2 describes the general pattern of ozone concentration distribution in the city. It clearly reflects that a number of sites with high concentration of the of the ozone such as Yateem Khana Chowk with 50 ppb, and Campus bridge with 86 ppb, and Charring cross with 90 ppb. Some sampling stations show low concentration of the ozone such as Kot Lakhpat with 15 ppb, Yadgar Chowk with 14 ppb, and Qartaba Chowk with 12 ppb. The high concentration of the ozone at the above stations may be due to the high traffic flow. At Yadgar Chowk its major cause of increase is the large number of vehicles passing through this station that is 74,897 (Fakhira, 1997). The increase may be due to some other reasons such as industries or congestion of the area.

Fig.4.3 describes the concentration into different zones which in fact reflect the changing patterns in the city; for example Band road and Badami Bagh comes under the category ranging from 20-40 ppb. While Campus Bridge and the Gulberg Main Market comes under the category ranging from 80-100 ppb. These zones are based on the factors such as housing density, rural and urban fringe and the traffic flow. Table 4.2 and 4.4 describe the patterns of change between 1993 and 2000. It is very much clear that concentration is changing from year to year. This change is clear from the bar graph, which describe the changing patterns of ozone in the city.

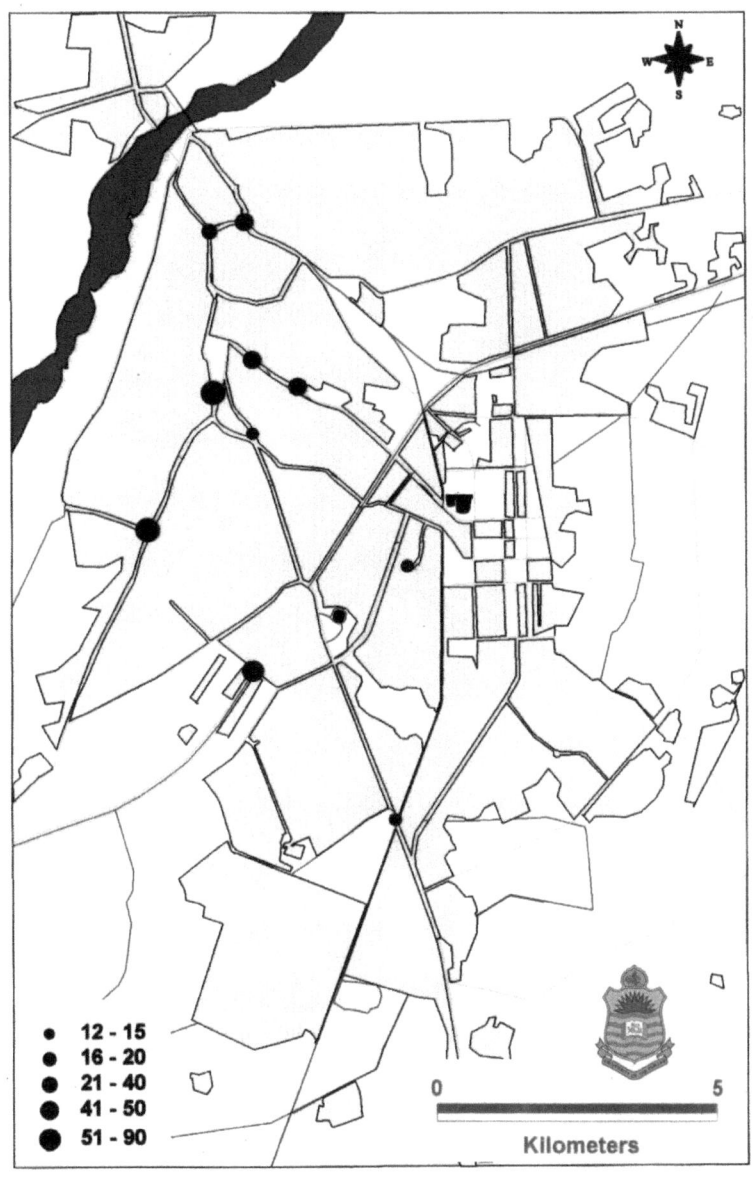

Figure 4.2: Patterns of Ozone in 2000 in ppb

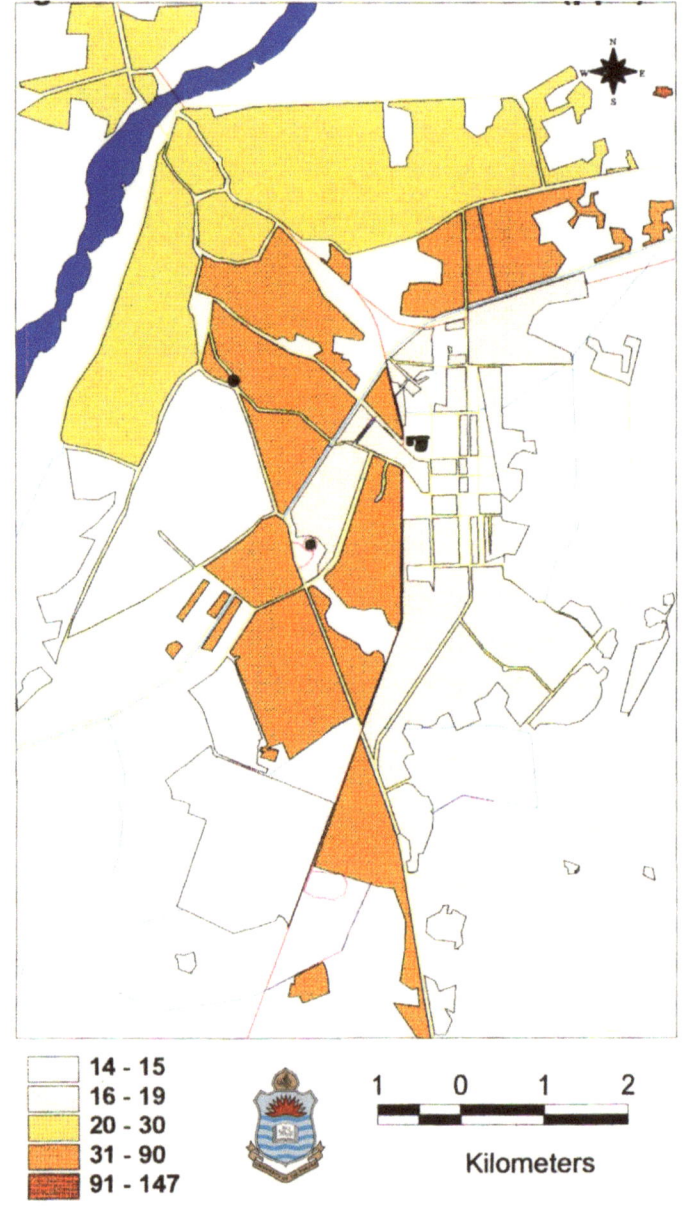

Figure 4.3: Patterns of Ozone in 2000 in ppb

Station	1993	1996	2000
Regal Chowk	0.7	39.2	35
EPA Chowk	7.1	17	20
Campus Bridge	9.6	10.8	86
Yateem Khana Chowk	3.1	18.5	50
Charring Cross	14.6	40	90
Yadgar Chowk	6.3	7.7	40
Gulberg Main Market	6	9.7	19
Qartaba Chowk	0.5	5	12
Badami Bagh Bus Station	21	27	30
Kot Lakhpat	10.2	10.7	15

Table 4.1: Patterns of Ozone (O_3) Concentration (in ppb)

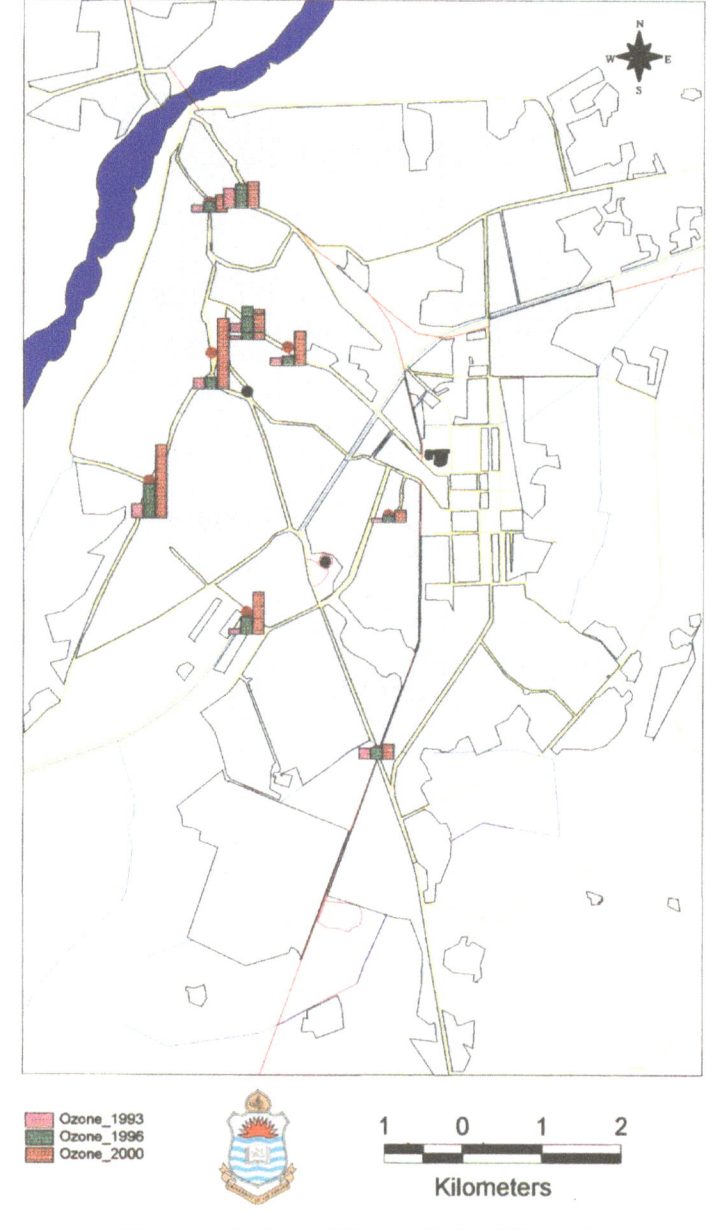

Figure 4.4: Patterns of Ozone 1993 - 2000 in ppb

4.3 Patterns of Sulphur Dioxide in the city in 2000

Sulphur dioxide in 2000 is 150 ppb at Yadgar Chowk. Its level is even higher than the level at which people experience bronchial spasms as described in the Chapter 2. According to the readings taken in 2000 concentration of the sulphur dioxide is higher at all the stations of Lahore taken as sampling points. Only at Campus Bridge its concentration is 20 ppb that is bearable.

Figure 4.6 shows that the concentration of sulphur dioxide in the different zones of the city. At Yadgar Chowk and in its surroundings the concentration of the ozone ranges from 141 to 160 ppb and at Yateem Khana Chowk and Charing Cross ranges from 81 to 100 ppb. There is very high concentration of ozone in this area because of high traffic flow. Gulberg and Qartaba Chowk fall under the category ranges from 21 to 40 ppb. In these areas concentration of sulphur dioxide is low because these are the open spaces. At Campus Bridge its concentration ranges from 0 to 20 ppb that is because of greenery it is low.

At Yateem Khana Chowk and Caring cross level of sulphur dioxide is 90 ppb which leads to the odour threshold level. At Mall road (Regal Chowk) sulphur dioxide level is 7.2 ppb in 1993 and 1996 it is 19.1 ppb. At Yateem Khana Chowk in 1996 the level of sulphur dioxide is 24.5 and in 2000 it is 90 ppb. An increase of 66 ppb can be seen in Figure 4.7. This increase is because of high traffic flow through this Chowk. At Campus Bridge in 1993 the concentration of sulphur dioxide was 1.9 ppb, which increased upto3.0 ppb in 1996. In 2000 its concentration of Sulphur dioxide is mainly because of traffic flow.

Figure 4.5: Patterns of Sulphur Dioxide in 2000 in ppb

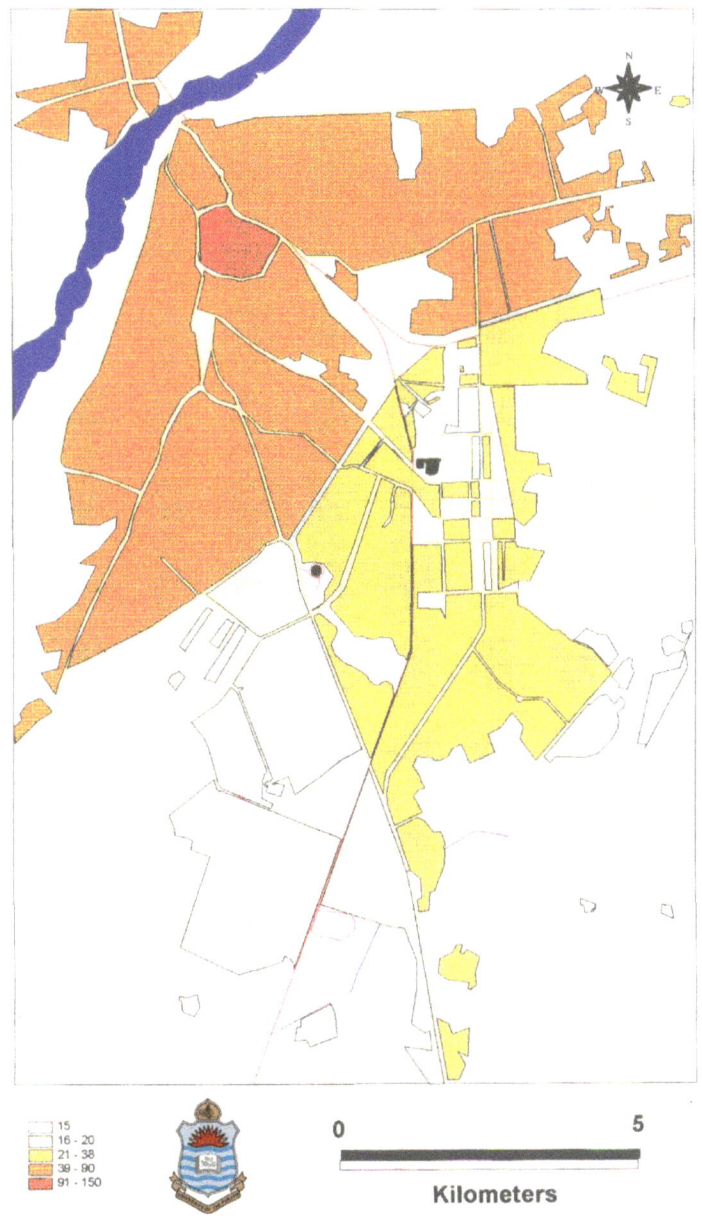

Figure 4.6: Patterns of Sulphur Dioxide in 2000 in ppb

Station	1993	1996	2000
Regal Chowk	7.2	19.2	80
EPA Chowk	3.0	3.1	65
Campus Bridge	1.9	3.0	20
Yateem Khana Chowk	13	24.5	90
Charing Cross	3.6	27.4	90
Yadgar Chowk	3.0	12.2	150
Gulberg Main Market	6.5	13.5	26
Qartaba Chowk	5.5	9.7	40
Badami Bagh Bus Station	10	22	60
Kot Lakhpat	3.7	10.5	50

Table 4.2: Patterns of Sulphur Dioxide (SO_2) concentration (in ppb)

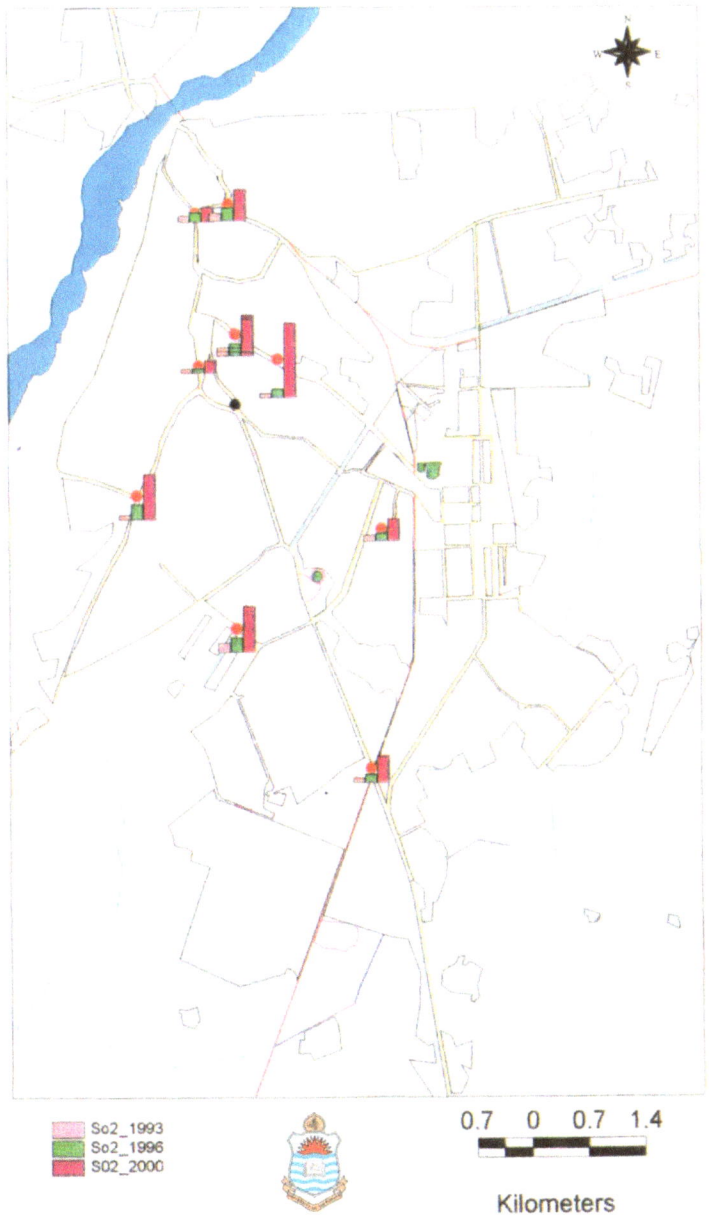

Figure 4.7: Patterns of Sulphur dioxide 1993-2000 (SO_2) conc. (in ppb)

4.4 Patterns of NO$_x$ in the city in 2000

Figure 4.10 describes the patterns of NO$_x$ concentration distribution in the Lahore city. It shows that at Charing Cross concentration of the NO$_x$ is 426 ppb and at Regal Chowk it is 300 ppb. Here concentration is very high because of high traffic flow. At Yadgar Chowk and Gulberg Main Market concentration of NO$_x$ is 225 and 210 ppb respectively because these are the industrial areas with high traffic flow. At Yadgar Chowk NO$_x$ concentration may be high because of other factors such as vertical expansion of the buildings. At both places Badami Bagh and Yateem Khana Chowk, its concentration is 179 because these are the industrial areas, very congested and with high traffic flow.

Fig.4.12 explains the patterns of NO$_x$ in the zones of Lahore city. At Charing Cross and its adjacent areas concentration of the NO$_x$ ranges from 401 to 500 ppb, while Regal Chowk, Yadgar Chowk, and Gulberg falls under the category ranges from 201 to 300 ppb. Campus Bridge, Yateem Khana Chowk, Badami Bagh, and Qartaba Chowk fall under the category that ranges from 101 to 200 ppb. Kot Lakhpat area has low concentration of NO$_x$ that ranges from 0 to 100 ppb.

In Figure 4.13 in 1993 NO$_x$ concentration is high at Yadgar Chowk. In 1996 it increased up to 158 and in 2000 225 ppb. The concentration of NO$_x$ is high at Yadgar Chowk because a huge flock of traffic passes through this Chowk. At Charing Cross in 1993 its concentration was 52 ppb, in 1996 114 ppb but in 2000 it is 426 ppb because according to a survey the motor vehicles passing through this Chowk in 24 hours are 74,493 (Fakhira, 1997).

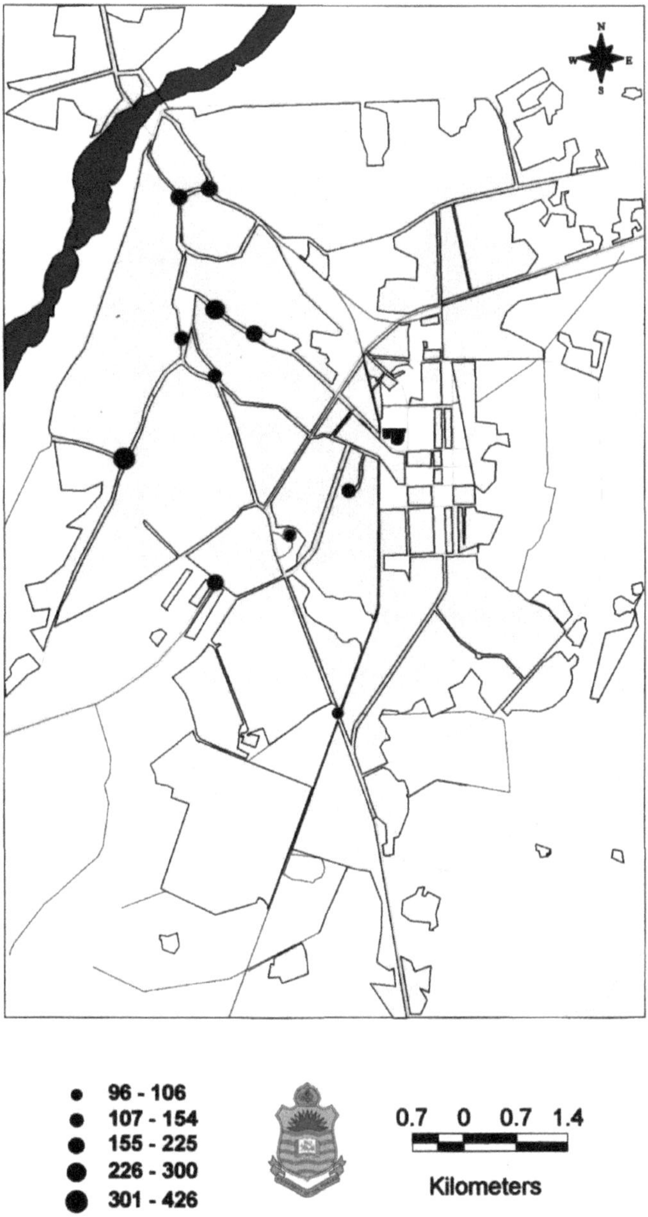

Figure 4.8: Patterns of NO_x in 2000 (in ppb)

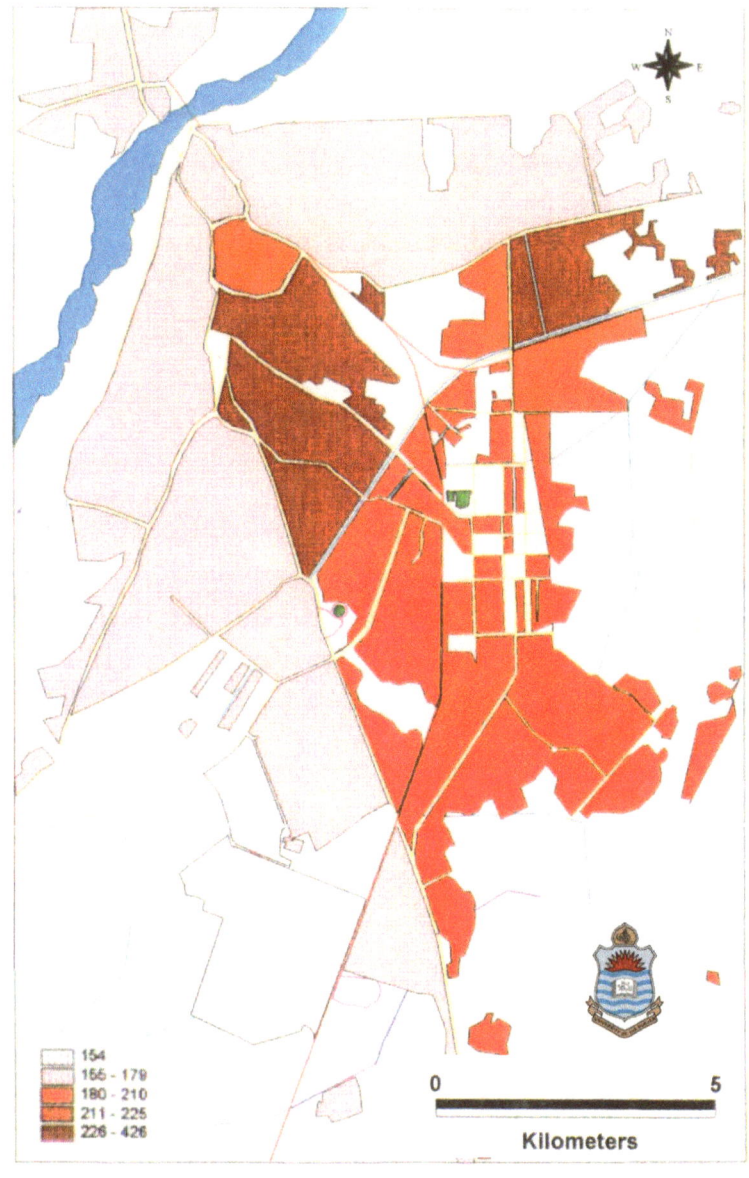

Figure 4.9: Patterns of NO_x in 2000 (in ppb)

Station	1993	1996	2000
Regal Chowk	35.1	129	300
EPA Chowk	22.1	23.6	106
Campus Bridge	4.3	17.3	154
Yateem Khana Chowk	86	91	179
Charring Cross	52	114	426
Yadgar Chowk	70	158	225
Gulberg Main Market	65	124	210
Qartaba Chowk	69	126	150
Badami Bagh Bus Station	37	58	179
Kot Lakhpat	3.7	27	96

Table 4.3: Patterns of NO_x concentration (in ppb)

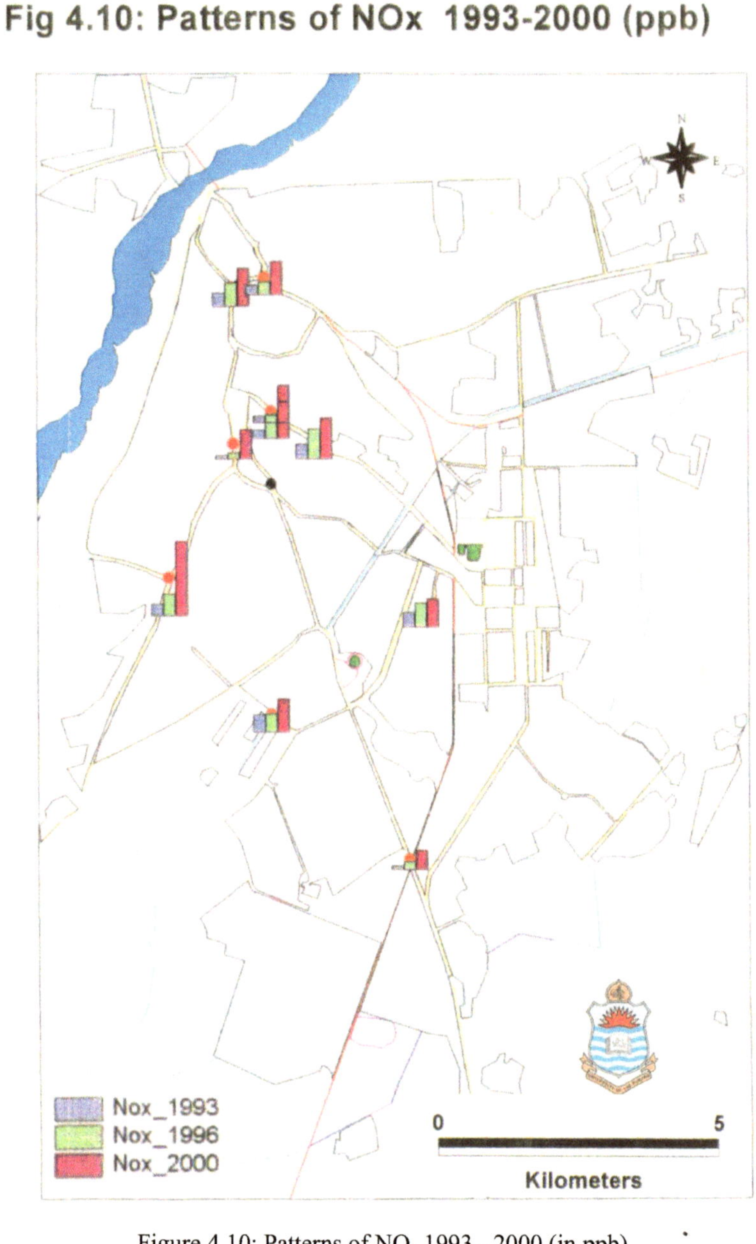

Figure 4.10: Patterns of NO_x 1993 - 2000 (in ppb)

4.5 Patterns of Carbon Monoxide in the City in 2000

Concentration of carbon monoxide at Badami Bagh and Qartaba Chowk is 7400 and 2200 ppb respectively as shown in the Figure 4.7. Concentration of carbon monoxide at Yadgar Chowk, Mall Road (Regal Chowk) and Charing Cross is 8000, 6400, 5000 respectively. Concentration of the carbon monoxide at the above stations is high enough to produce headache, and train on the heart. As described in the Figure 4.6 the concentration of the carbon monoxide falls between 7 to 8 ppb at the Yadgar Chowk and its adjoining areas because of high traffic flow and the vertical development of the buildings. At Mall Road concentration of the carbon monoxide ranges from 5000 to 6000. At Gulberg and the adjoining areas concentration of carbon monoxide falls between 3000 to 4000 ppb because these areas are commercial as well as industrial.

In 1993 at Mall road (Regal Chowk), concentration of the carbon monoxide is 2700 ppb while in 1996 its concentration is 3800 ppb. So an increase of 1100 ppb can be seen in the Figure 4.9. In year 1993, its concentration is 4700 ppb and in year 2000 it is 6400 ppb. It has been stated in Chapter 2 that major source of carbon monoxide is the motor vehicle, so an increase in the traffic led to the increase in the carbon monoxide level. At Gulberg and Campus its concentration is 4100 and 3600 ppb because these are open spaces with the green belts.

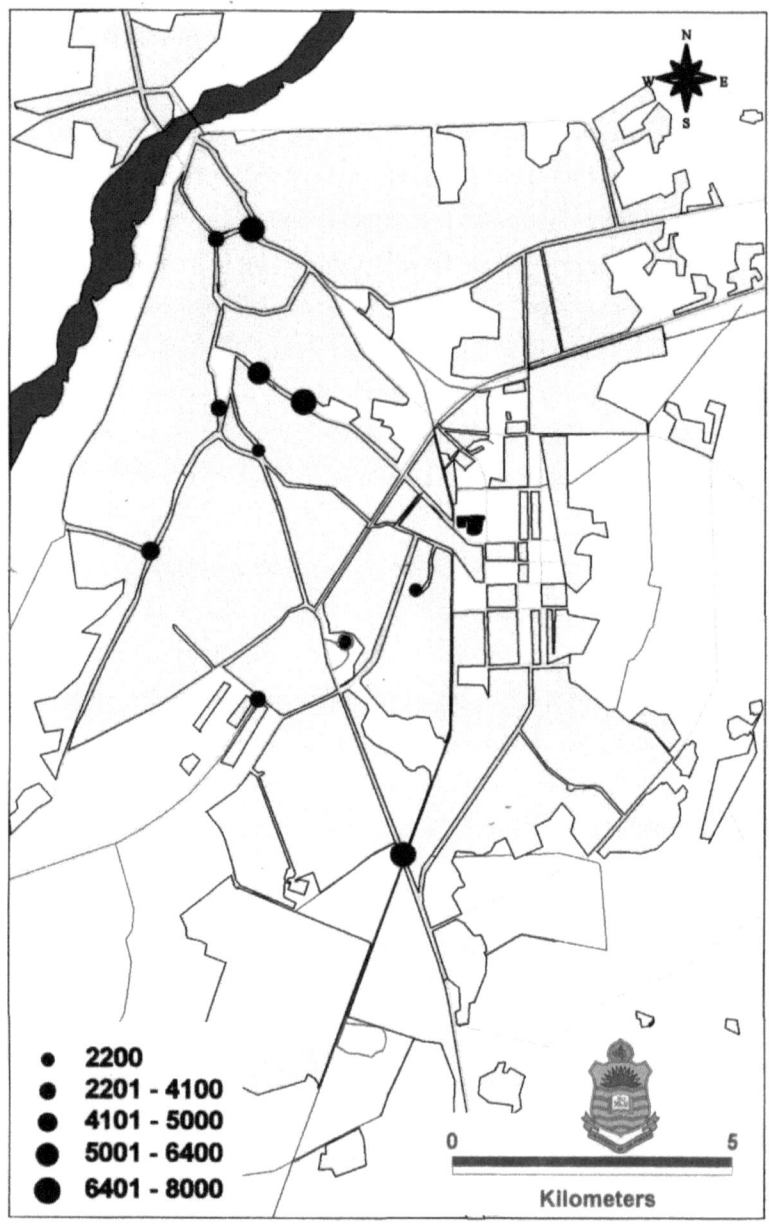

Figure 4.11: Patterns of carbon monoxide in 2000 (in ppb)

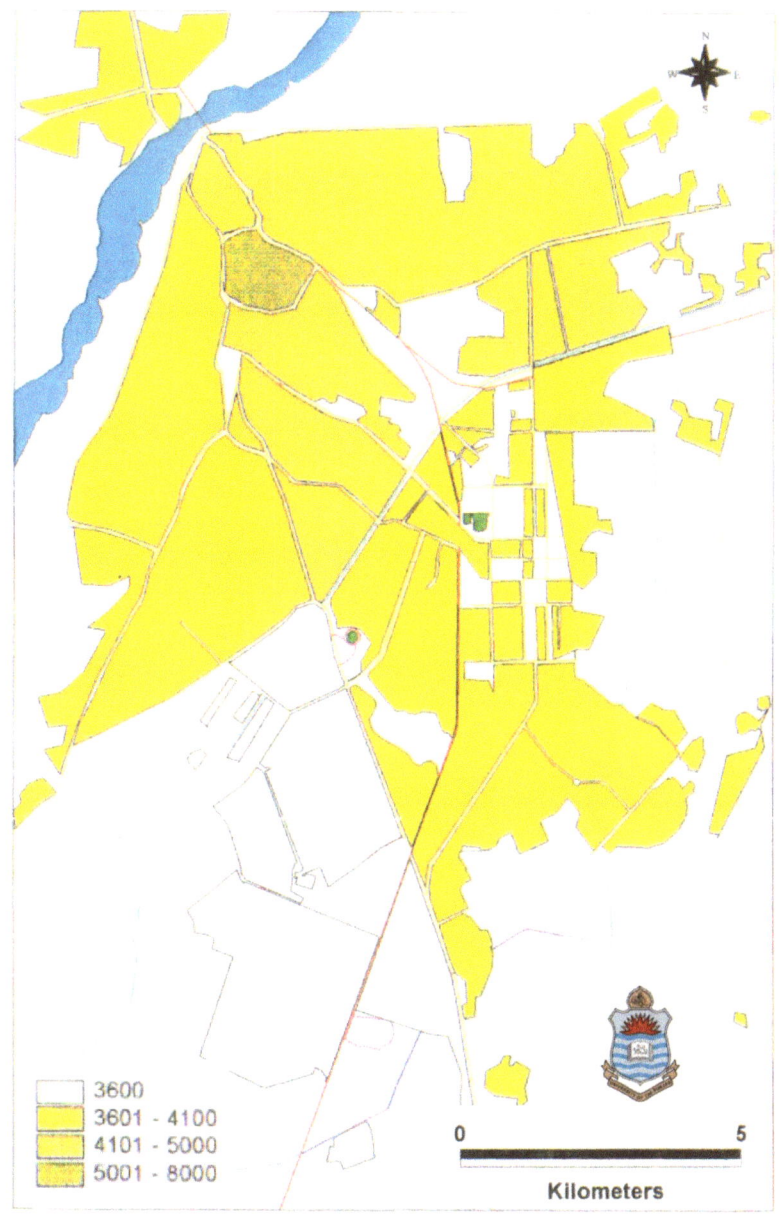

Figure 4.12: Patterns of carbon monoxide in 2000 (in ppb)

Station	1993	1996	2000
Regal Chowk	2700	3800	6400
EPA Chowk	800	3200	6000
Campus Bridge	800	900	154
Yateem Khana Chowk	600	1900	4100
Charring Cross	400	5200	5000
Yadgar Chowk	450	2300	8000
Gulberg Main Market	390	1100	4100
Qartaba Chowk	700	1200	2200
Badami Bagh Bus Station	3500	3900	7400
Kot Lakhpat	1260	4000	7900

Table 4.4: Patterns of carbon monoxide (CO) concentration (in ppb)

Figure 4.13: Patterns of carbon monoxide 1993 – 1996 – 2000 (in ppb)

4.6. Patterns of Particulate Matter in the city in 2000

Figure 4.14 shows that concentration of particulate matter at Campus Bridge is 1165. It may be high due to open space because with more sunshine in the open space level of particulate matter becomes high and they remain suspended in the air for long time. At Yateem Khana Chowk concentration of the particulate matter is 1123 micrograms per cubic metres because the motor vehicles passing through this Chowk are 21492, which represents a high traffic volume. At Regal Chowk and Charring Cross concentration of the particulate matter is 1100 and 1050 respectively. Concentration of the particulate matter is 980 and 890 at EPA Chowk and Gulberg respectively.

Figure 4.14 describes the patterns of particulate matter in different zones of the Lahore city. Campus Bridge, Yateem Khana Chowk and the Qartaba Chowk have concentration of particulate matter ranging from 1000 to 1200 and except Yadgar Chowk all the stations fall in the range from 801 to 1000.

In Figure 4.16 the changing patterns of particulate matter in the city have been shown. In year 1993 its concentration at Campus Bridge is 290 and in year 1996 it is 684, while in year 2000 it is 1165. Its concentration increased due to high traffic flow. At Yateem Khana Chowk it is 639 in year 1993, 913 in year 1996, and 1123 in year 2000 because it is an industrial as well as area with high traffic flow.

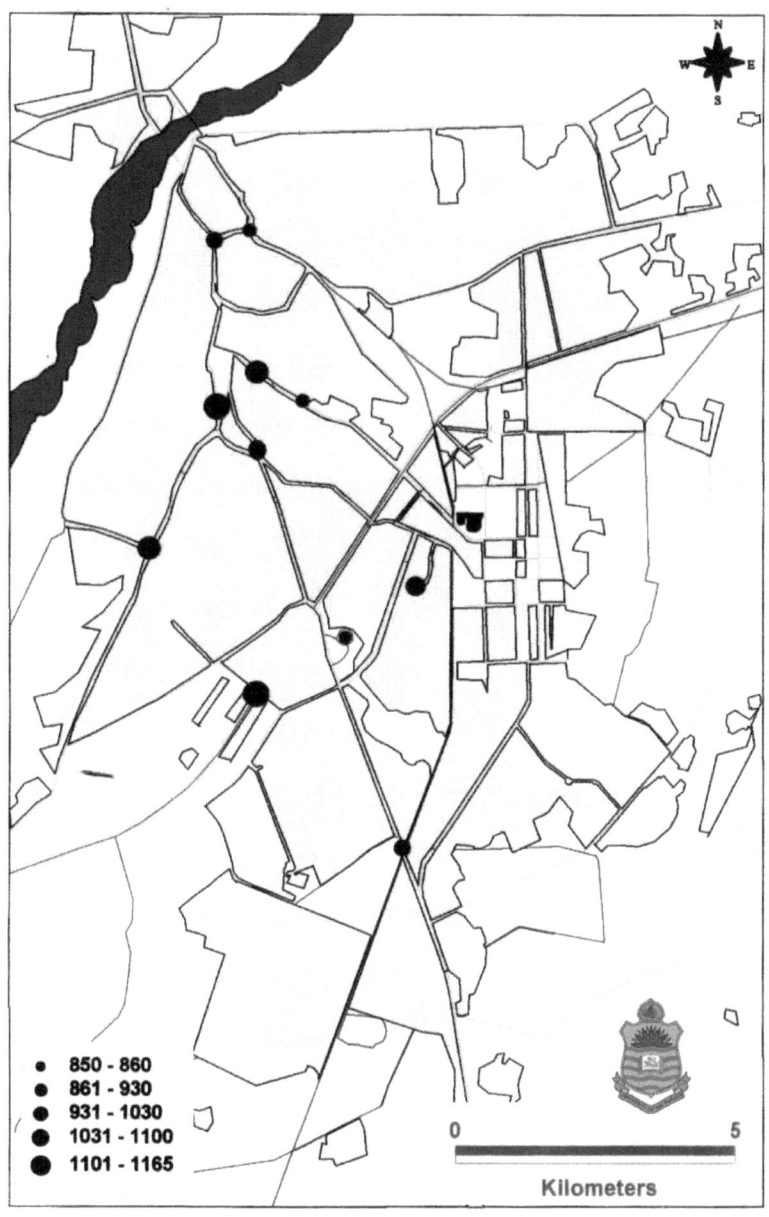

Figure 4.14: Patterns of Particulate Matter in 2000 (microgram per cubic meter)

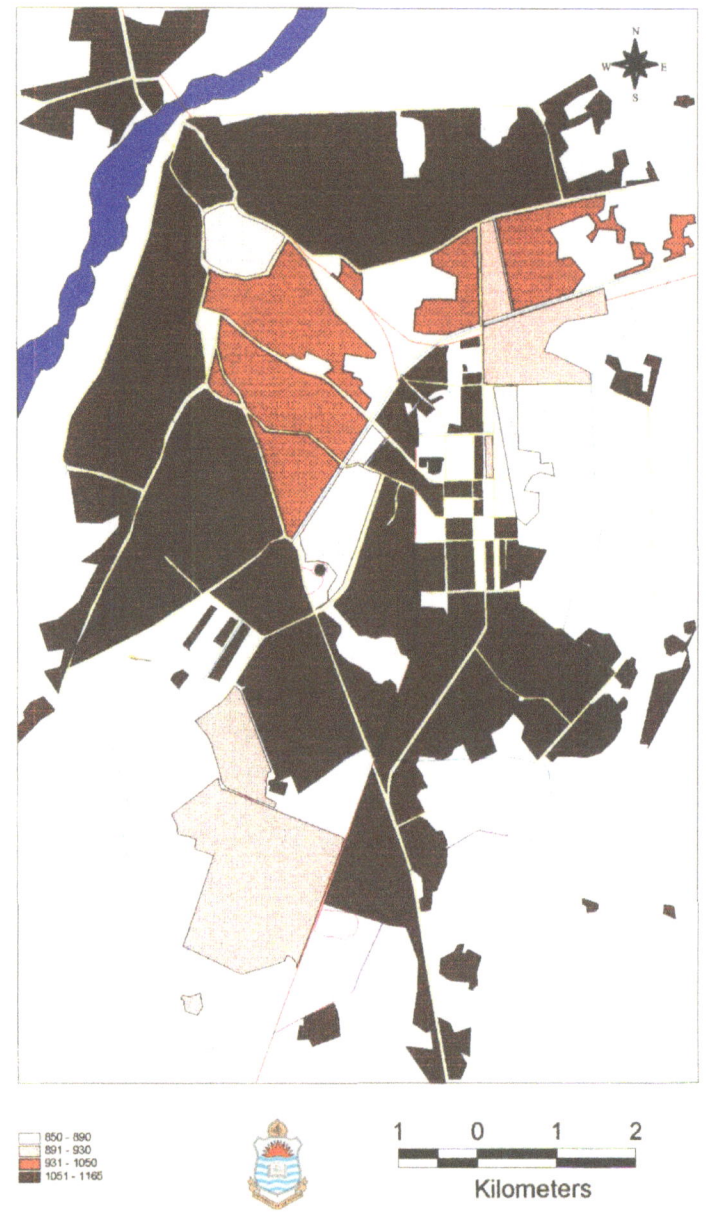

Figure 4.15: Patterns of particulate Matter in 2000 (microgram per cubic meter)

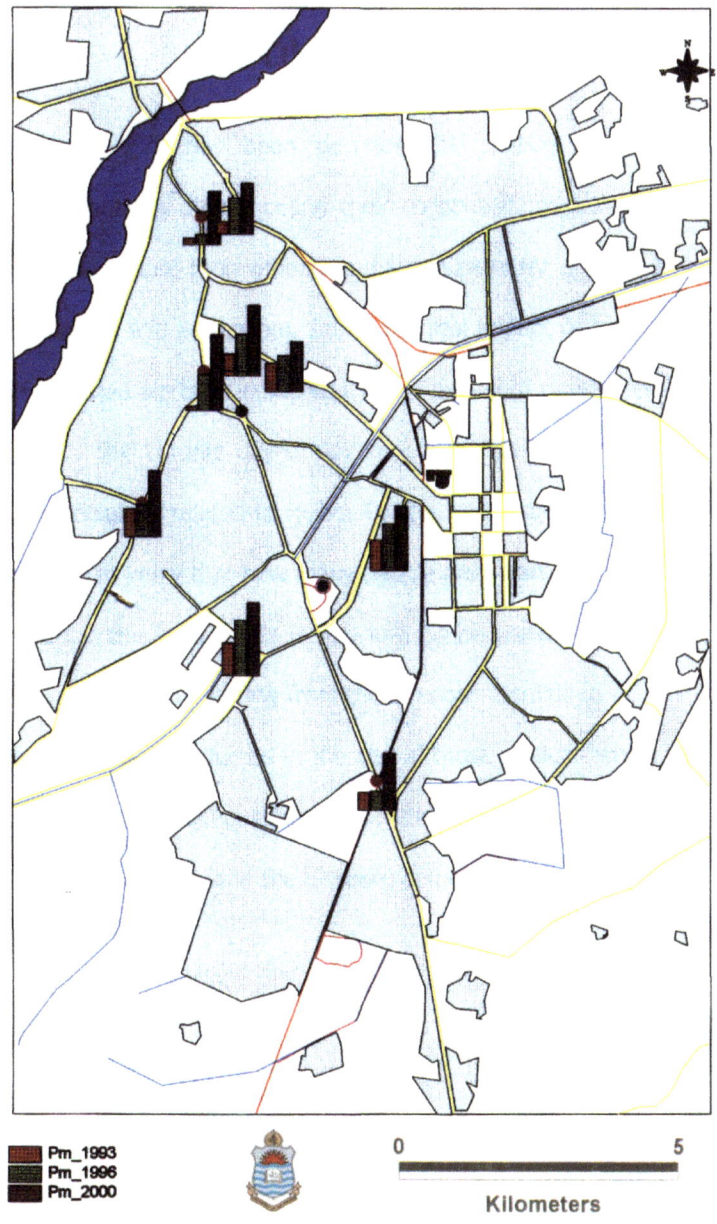

Figure 4.16: Patterns of particulate Matter 1993 – 1996 – 2000

Station	1993	1996	2000
Regal Chowk	390	465	1100
EPA Chowk	450	690	980
Campus Bridge	290	684	1165
Yateem Khana Chowk	639	913	1123
Charring Cross	587	630	1050
Yadgar Chowk	590	682	850
Gulberg Main Market	320	370	890
Qartaba Chowk	619	817	1030
Badami Bagh Bus Station	400	682	860
Kot Lakhpat	455	473	930

Table 4.5: Patterns of Particulate Matter (microgram per cubic meter)

Chapter 5:
People's Behaviour towards Air Pollution

5.1 Public survey

In Chapter 2, it has been described that effects of pollutants vary considerably because of difference in their concentrations and their chemistry. Some are more toxic than others, and some have far greater impacts than others on materials and ecosystem.

In fact the real impact of air pollution is not so much the isolated alerts or incidents that kill people or make them ill for a day or two, as it is the chronic day-to-day pollution that leads to lung diseases, headache, and diseases related to vision. To control the air pollution problem it is the need of hour to know that how many people are aware of this problem.

In Pakistan the literacy rate is very low. So people don't take it as a major problem though they are suffering from the diseases mentioned above.

This survey was conducted in the city at those stations which are thought to be more polluted. This questionnaire reflects that how much people are affected by the air pollution and the diseases caused by the air pollution.

S. No.	ID	Affected	Not affected
1	Campus Bridge	100	0
2	Gulberg Main Market	60	40
3	Mall Road [GPO]	90	0
4	Anarkali	100	0
5	Railway Station	100	0
6	Badami Bagh	100	0
7	Yateem Khana Chowk	100	0
8	Sarwar Road Cantt.	100	0
9	Model Town Residencial Area	100	0
10	Defence Y-Sector	100	0

Table 5.1   People's awareness about the effects of air pollution

Many people think that they are not affected by the pollution because of an increase in the illiteracy rate. They don't even know that how much dangerous it is.

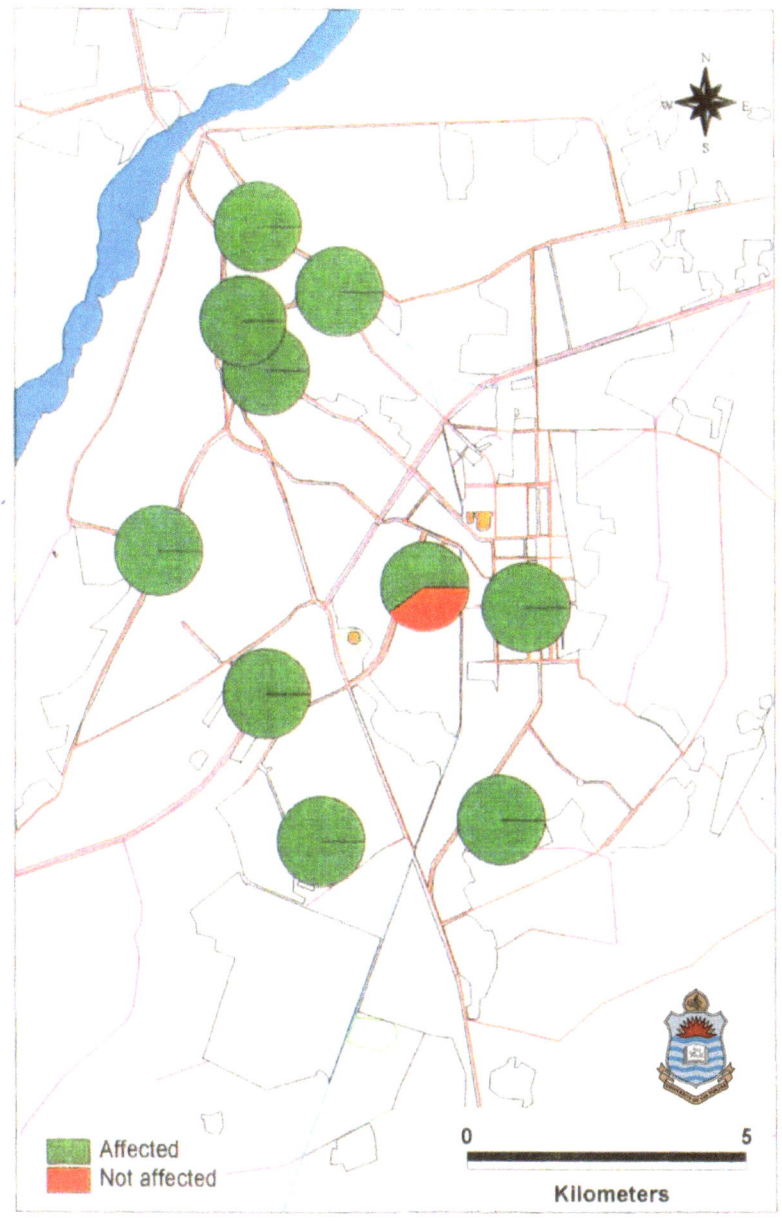

Figure 5.1: People awareness about the effects of air pollution

S. No.	ID	Affected	Not affected
1	Campus Bridge	80	20
2.	Gulberg Main Market	90	10
3.	Mall Road [GPO]	70	30
4.	Anarkali	40	60
5.	Railway Station	0	100
6.	Badami Bagh	0	100
7.	Yateem Khana Chowk	50	50
8.	Sarwar Road Cantt.	90	10
9.	Model Town residential Area	70	30
10.	Defence Y-Sector	90	10

Table 5.2 People's awareness towards Government's responsibility about pollution

Many people think that govt. is not taking any step to control the pollution as government should implement some laws to control it.

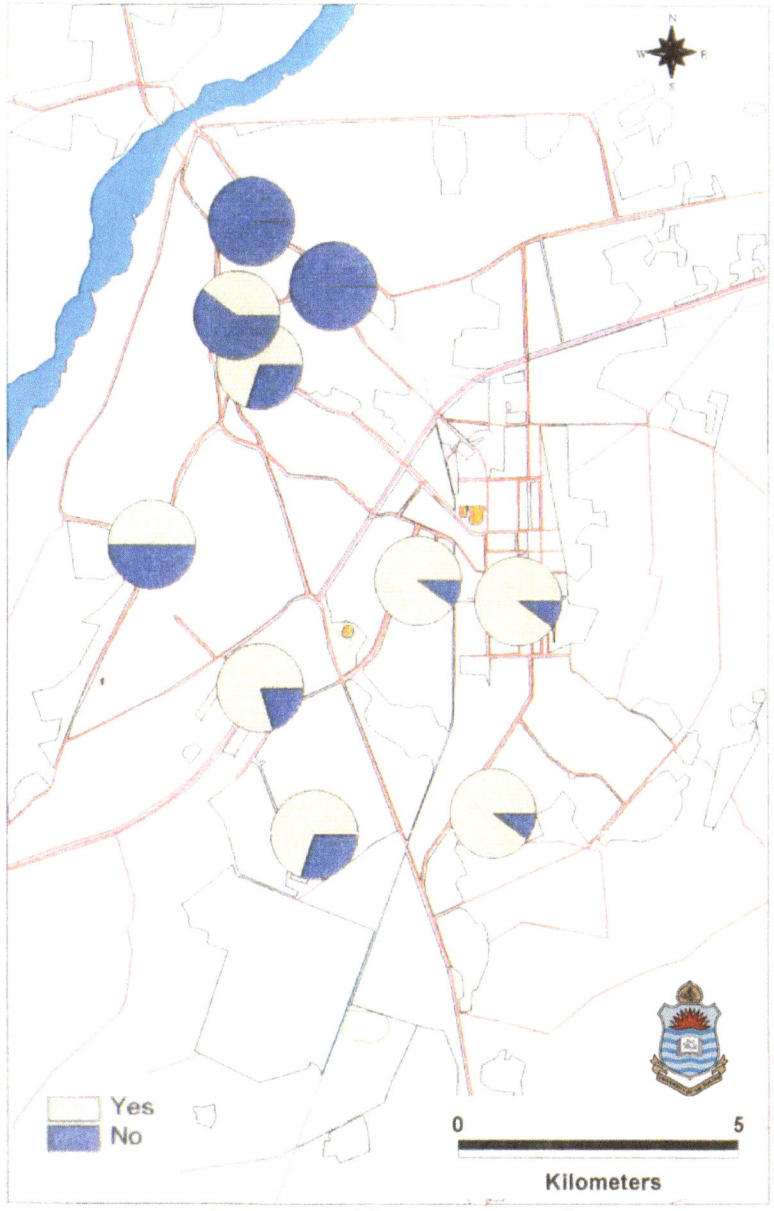

Figure 5.2 People's awareness towards
Government's responsibility about pollution

S. No.	ID	Affected	Not affected
1	Campus Bridge	90	10
2.	Gulberg Main Market	50	50
3.	Mall Road [GPO]	40	60
4.	Anarkali	40	60
5.	Railway Station	0	100
6.	Badami Bagh	0	100
7.	Yateem Khana Chowk	30	70
8.	Sarwar Road Cantt.	70	30
9.	Model Town Residential Area	65	35
10.	Defence Y-Sector	70	30

Table 5.3: People's behaviour towards control of air pollution

A high percentage is doing nothing to control the air pollution even they can't keep their houses and shops clean properly

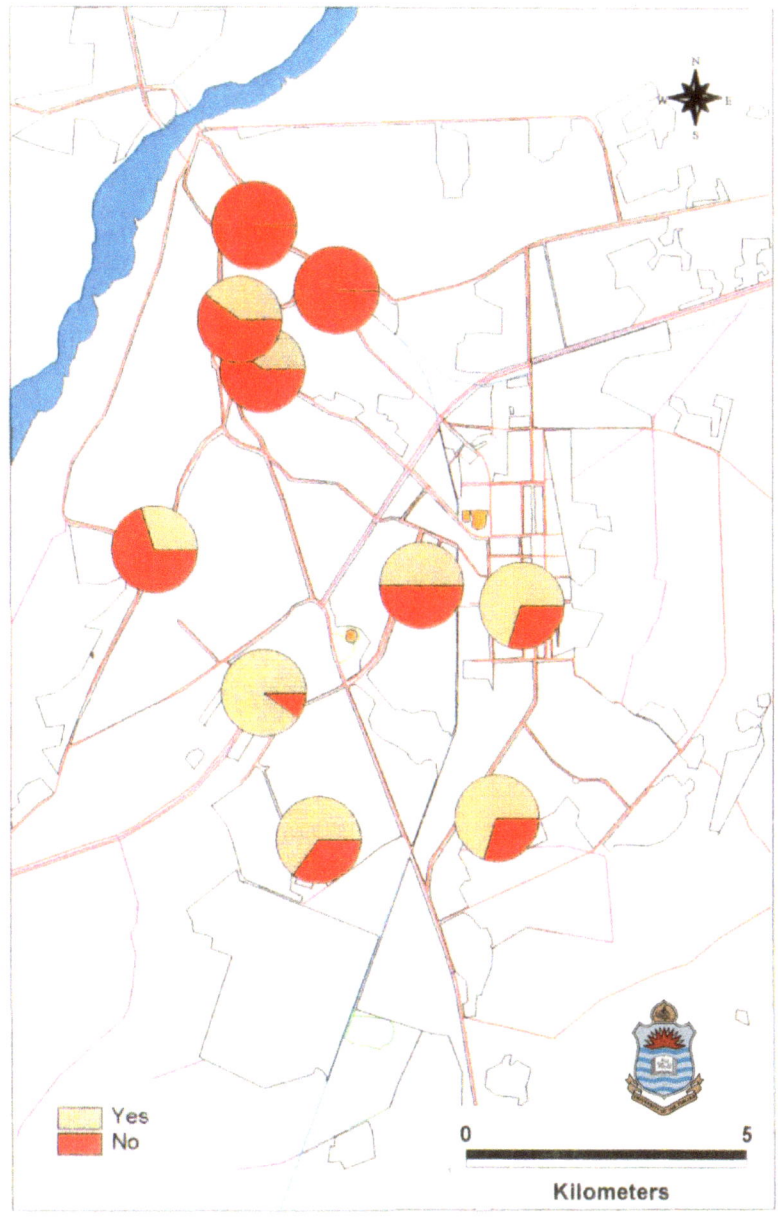

Figure 5.3: People's behaviour towards control of air pollution

S. No.	ID	Affected	Not affected
1	Campus Bridge	0	100
2.	Gulberg Main Market	0	100
3.	Mall Road [GPO]	2	98
4.	Anarkali	0	100
5.	Railway Station	0	100
6.	Badami Bagh	0	100
7.	Yateem Khana Chowk	0	100
8.	Sarwar Road Cantt.	0	100
9.	Model Town Residential Area	0	100
10.	Defence Y-Sector	0	100

Table 5.4 People's awareness about environment protection department

ETA is not very famous in Lahore because they don't display any board as well as cautions on the roads.

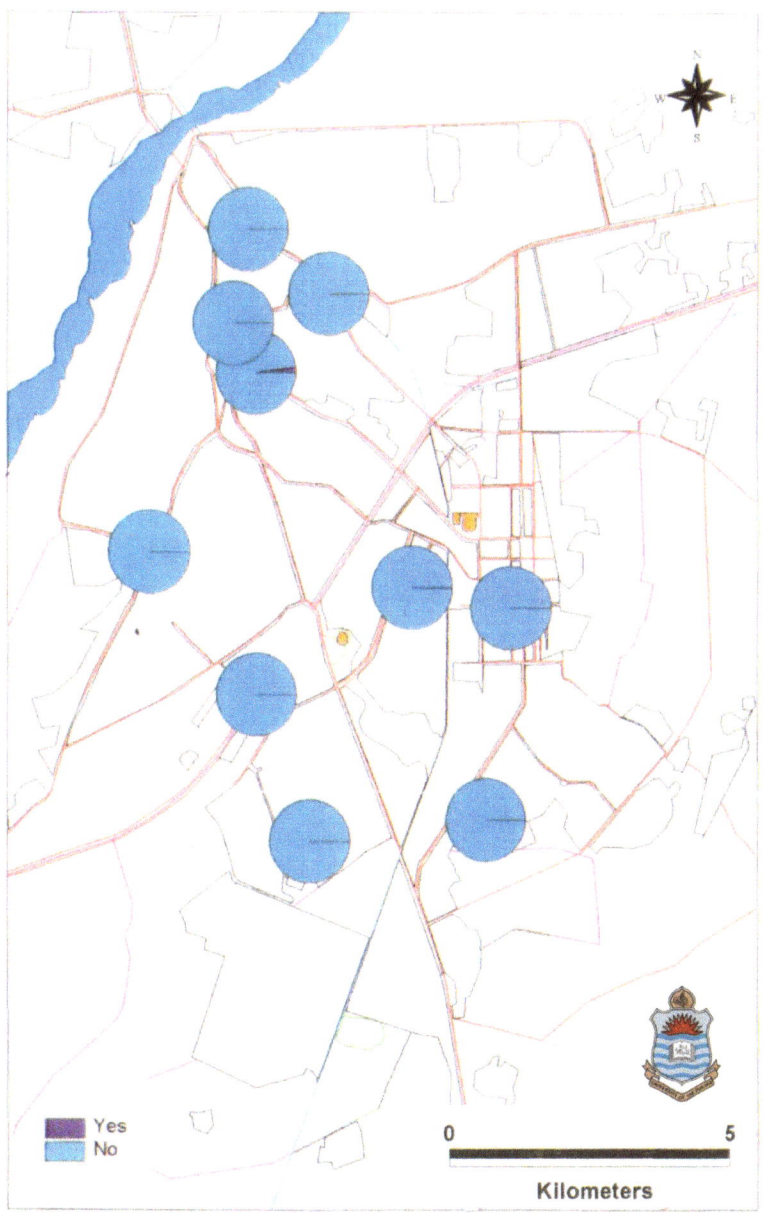

Figure 5.4 People's awareness about environment protection department

S. No.	ID	Affected	Not affected
1	Campus Bridge	100	0
2.	Gulberg Main Market	90	10
3.	Mall Road [GPO]	100	0
4.	Anarkali	100	0
5.	Railway Station	100	0
6.	Badami Bagh	80	20
7.	Yateem Khana Chowk	100	0
8.	Sarwar Road Cantt.	100	0
9.	Model Town Residential Area	100	0
10.	Defence Y-Sector	100	0

Table 5.5 Attitude towards effects of pollution on the health

Different diseases are caused by the pollution, so many people know that pollution is affecting their health

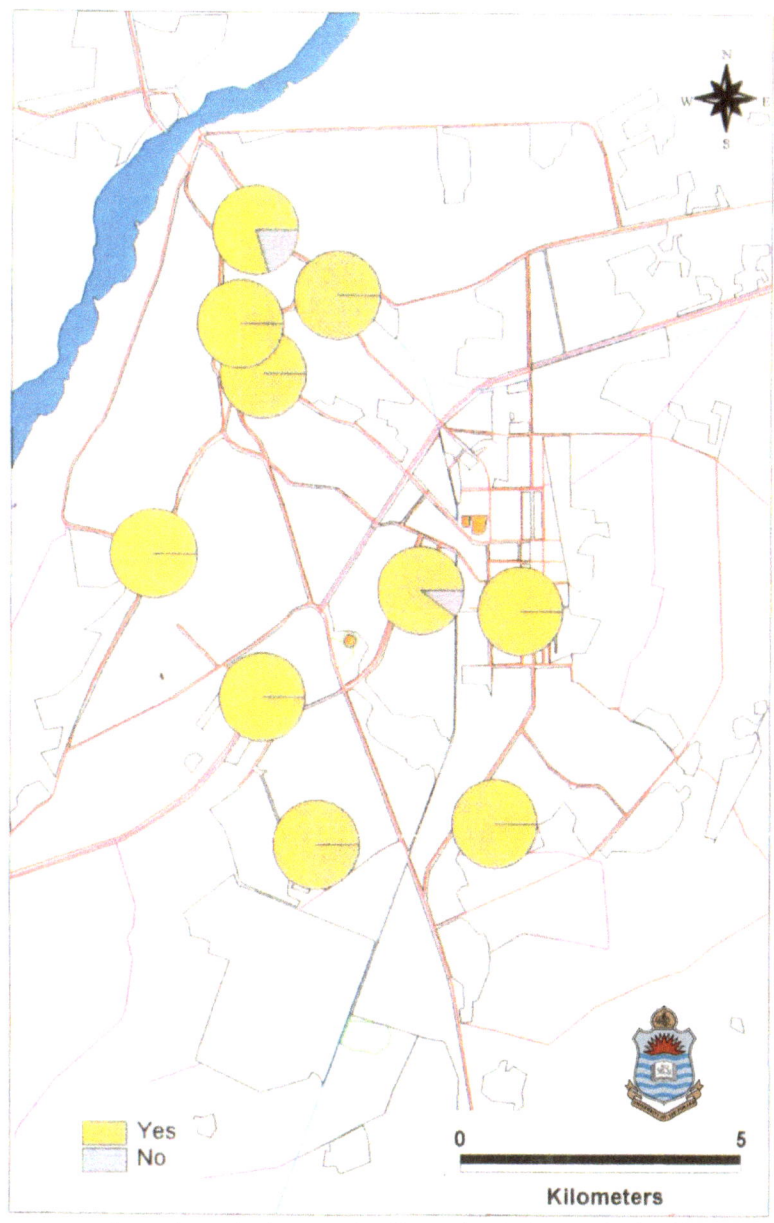

Figure 5.5 Attitude towards effects of pollution on the health

S. No.	ID	Affected	Not affected
1	Campus Bridge	100	0
2.	Gulberg Main Market	80	20
3.	Mall Road [GPO]	100	0
4.	Anarkali	90	10
5.	Railway Station	100	0
6.	Badami Bagh	100	0
7.	Yateem Khana Chowk	100	0
8.	Sarwar Road Cantt.	100	0
9.	Model Town Residential Area	100	0
10.	Defence Y-Sector	100	0

Table 5.6 Attitude towards motor vhicles

In America, the automobile sources like diesel-powered trains and trucks and planes constitutes at least 42% of the major five air pollutants. Pakistan data is not available in this regard yet it is estimated that Pakistan may have similar figures. Automobiles emit primary and secondary pollutants in air through combustion

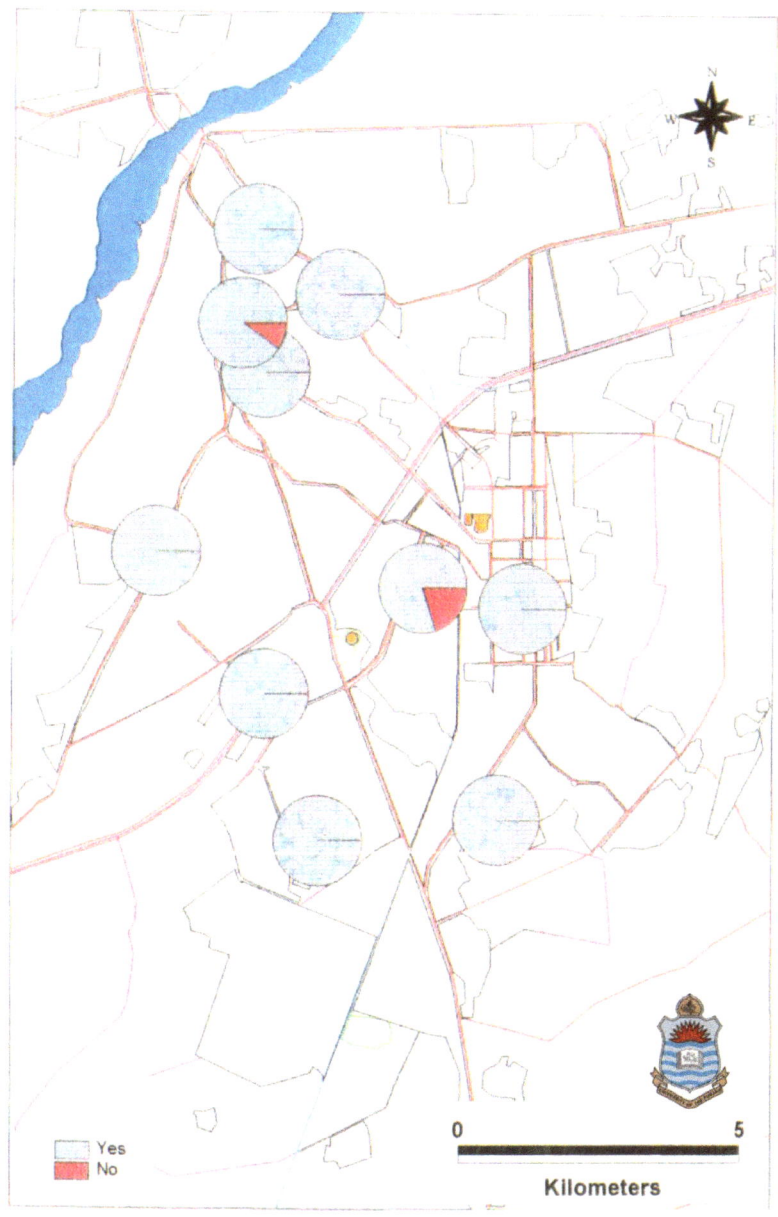

Figure 5.6 Attitude towards motor vehicles

S. No.	ID	CHILD%	YOUNG %	OLD%
1	Campus Bridge	45	20	35
2.	Gulberg Main Market	34	22	44
3.	Mall Road [GPO]	60	10	30
4.	Anarkali	40	20	40
5.	Railway Station	10	60	30
6.	Badami Bagh	20	70	10
7.	Yateem Khana Chowk	25	65	10
8.	Sarwar Road Cantt.	60	10	30
9.	Model Town Residential Area	34	8	64
10.	Defence Y-Sector	50	20	30

Table 5.7 Age wise effects of air pollution on people

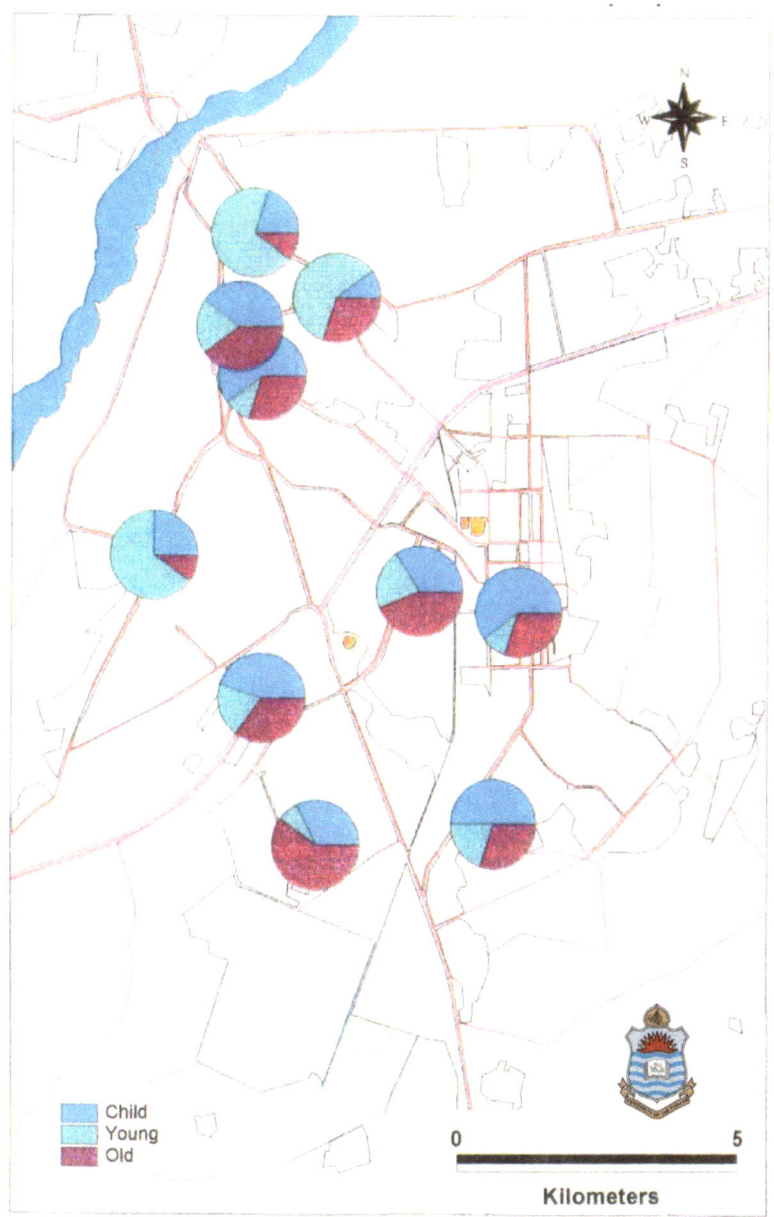

Figure 5.7 Age wise effects of air pollution on people

S. No.	ID	RICKSHAW	WAGON	OTHER AUTOMOBILES
1	Campus Bridge	40	20	40
2.	Gulberg Main Market	20	0	80
3.	Mall Road [GPO]	70	30	0
4.	Anarkali	80	15	5
5.	Railway Station	90	10	0
6.	Badami Bagh	100	0	0
7.	Yateem Khana Chowk	80	0	20
8.	Sarwar Road Cantt.	45	10	45
9.	Model Town Residential Area	70	30	0
10.	Defence Y-Sector	100	0	0

Table 5.8 Opinion towards major source of pollution

Rickshaw and wagons have two-stroke, so they produce more pollution.

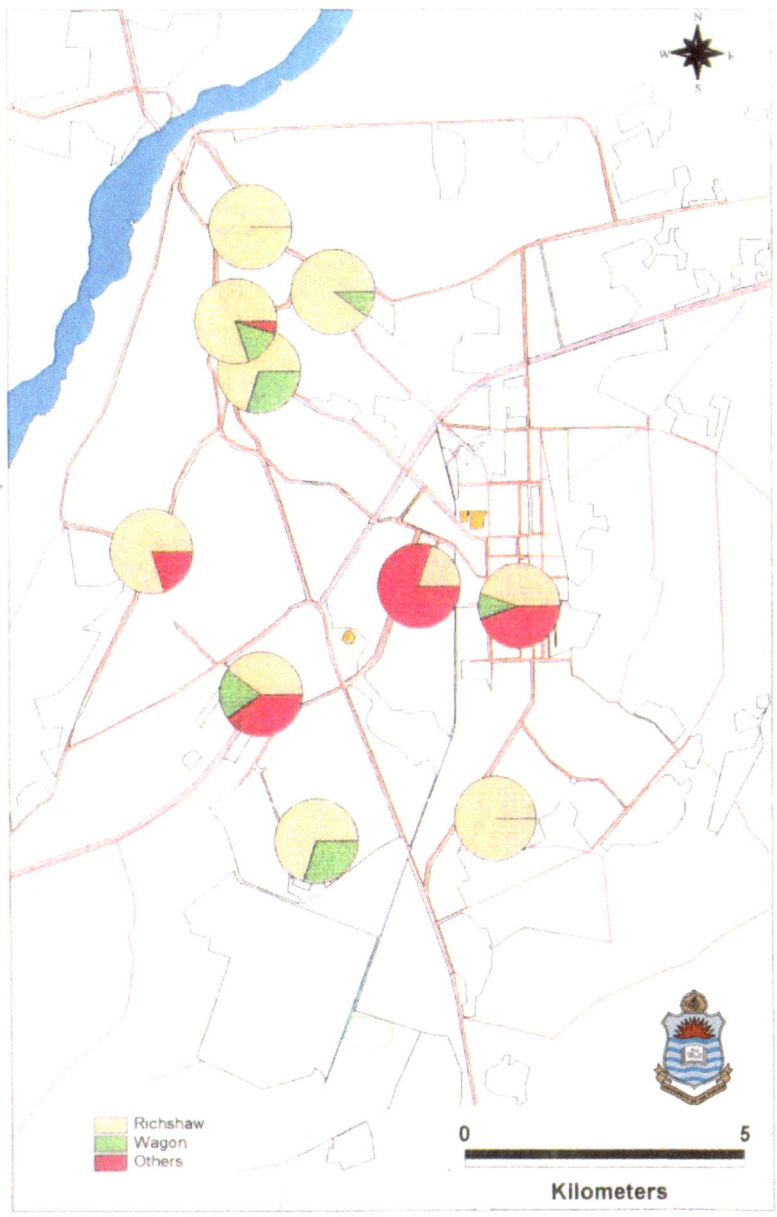

Figure 5.8 Attitude towards major source of pollution

S. No.	ID	Eye irritation	Headache	Nausea	Lung problem
1	Campus Bridge	40	20	30	10
2.	Gulberg Main Market	20	60	10	10
3.	Mall Road [GPO]	70	20	0	10
4.	Anarkali	40	30	0	30
5.	Railway Station	80	0	0	20
6.	Badami Bagh	40	40	19	1
7.	Yateem Khana Chowk	89	10	0	1
8.	Sarwar Road Cantt.	40	30	10	20
9.	Model Town Residential Area	20	60	15	5
10.	Defence Y-Sector	10	70	15	6

Table 5.9 Health problems faces by people

This report shows that, How many people are affected by the pollution and the nature of the diseases caused by the pollution

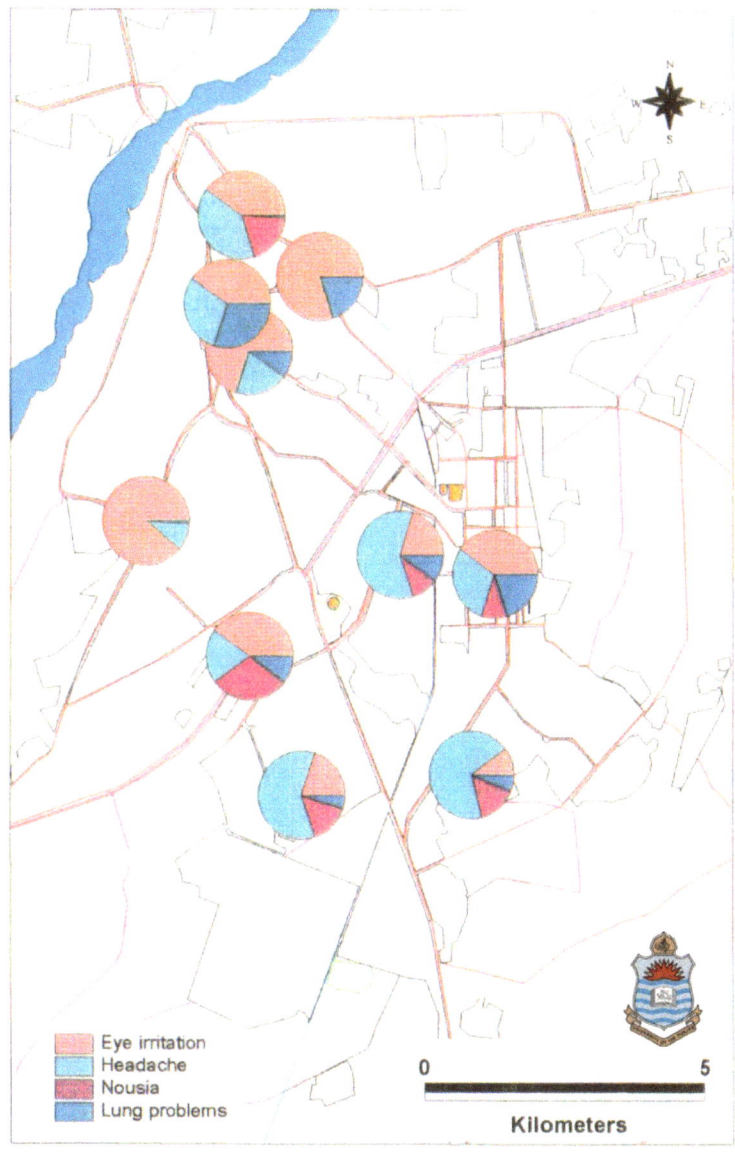

Figure 5.9 Health problems faces by people

S. No.	ID	YES	NO
1	Campus Bridge	100	0
2.	Gulberg Main Market	60	40
3.	Mall Road [GPO]	60	40
4.	Anarkali	70	30
5.	Railway Station	0	100
6.	Badami Bagh	0	100
7.	Yateem Khana Chowk	40	60
8.	Sarwar Road Cantt.	65	35
9.	Model Town Residential Area	90	10
10.	Defence Y-Sector	100	0

Table 5.10 Solution pf problem

Most of the people think that pollution is such a major problem that cannot be overcome by a developing country.

Figure 5.10 Solution pf problems

Left blank for pagination

Chapter 6:
Summary and Conclusion

6.1 Summary

The spatial patterns of air pollution in Lahore City reflect that the more emphasis in the present study has been placed on the present day patterns of air pollution. For this purpose ten sampling points were selected and the changing pattern of pollutants i.e. ozone, sulphur dioxide, carbon monoxide, oxides of nitrogen, and particulate matter had been studied. In the very first chapter introduction of the air pollution and its understanding on the global as well as country and regional level has been discussed. Data sources and methodology have been described in this chapter.

In the second part, air pollutants, their sources and effects have been discussed which not only reflects the nature of pollutants but also describes the diseases caused by the high or low concentration of these pollutants.

The third chapter provides the detail information about the Lahore City. In this part emphasis has been given on the climate including temperature, rainfall, humidity and wind direction, and population growth of the city along with the existing land use.

A survey in fifth part reflects the behaviour of the people towards air pollution & the diseases caused by air pollution and the awareness among people. This survey also helps in the preventive and control measures.

6.2 Conclusion

The results of the study show that concentrations of all the pollutants are increasing day by day. Such as the Table 4.1 reflects the changing pattern of ozone from 1993-2000 as well as in 2000 at different places. Ozone concentration is high at Charing Cross, i.e., 19 ppb. This is high enough to cause lung diseases as well as the vision problems. At Campus Bridge its concentration in year 2000 is 86 ppb, which raised rapidly from year 1993 to year 2000.

In Table 4.2 changing pattern of SO_2 are shown. Its concentration is 150 ppb at Yadgar Chowk, being an industrial as well as congested area. Its concentration is also high at Regal Chowk, Yateem Khana and Charing Cross as shown in Table 4.2

At Campus Bridge its concentration is 20 ppb which is low probably due to vegetation. Table 4.3 reflects the concentration patterns of NO_x in the city. Its concentration is 426 ppb at Charing Cross and 225 ppb at Yadgar Chowk because the former one is a commercial area while the later one is in industrial area.

Carbon monoxide concentration is 8000 ppb at Yadgar Chowk & leading to Regal Chowk & Charing Cross with 6400 ppb & 5000 ppb respectively.

At Yadgar Chowk its level raised very rapidly from year 1993 to year 2000. Concentration of particulate matter is 1165 at Campus Bridge & raised very rapidly from 290 to 1165 ug/m^3 at campus from 1993 to 2000.

Fourth part of the study shows the spatial pattern of air pollution at ten sampling sites. The changing patterns of concentration of each gas at these sampling sites have been shown on the map.

The maps with the dot symbol show the concentration of each gas in year 2000 at each point, which are also comparable with other points. The city has been divided into different zones on the bases of intensity of gases more over temporal changes has been shown since year 1993, moreover time based variation also examined at sampling sites.

An overall pattern of increase in all the pollutants reflects that Charing Cross and Yadgar Chowk are the most polluted areas with high concentrations of all the pollutants.

6.3 Recommendations:

Govt. should implement some laws to control the air pollution such as 1). Industries should be set up away from the city centre & the residential areas, 2). Some motor vehicular emission standards should be implemented by the government, 3). More and more vegetation should be planted, 4). Awareness of air pollutants and their effects should me made among people through media.

Left blank for pagination

Annex-1:
Bibliography

Anderson, H. R. (1999) [Editors Holgate, S. T., H. S. Koren, et al. (1999)]. Health effects of air pollution episodes [*in "Air pollution and health"*], Academic Press. ISBN: 9780080526928. 1065pp.

Barker, J. R. and D. T. Tingey (1992). Air-pollution effects on biodiversity. Other Information: ManTech Environmental Technology, Inc., Corvallis, OR: 328pp.

Bruce, N., Perez-Padilla, R., & Albalak, R. (2000). Indoor air pollution in developing countries: a major environmental and public health challenge. *Bulletin of the World Health Organization*, 78(9), 1078-1092.

Dockery, D. W., & Pope, C. A. (1994). Acute respiratory effects of particulate air pollution. *Annual review of public health*, 15(1), 107-132.

Fakhira Zahir (1997). *Spatial Analysis of Transport Network in Lahore City*. Department of Geography, University of the Punjab. MSc Thesis.

Khan, M. H., & Siddiqui, A. S. (1982). Growth and Fluctuations in the Output of Major Crops in Pakistan, 1950-51 to 1979-80. *The Pakistan Development Review*, 149-158.

Kupchella, C. E. and M. C. Hyland (1989). *Environmental Science: Living Within the System of Nature*, 2nd Edition. Allyn and Bacon. ISBN: 0205120164, 9780205120161. 637pp

Kupchella, C. E. and M. C. Hyland (1993). *Environmental Science: Living Within the System of Nature*, 3rd Edition. Prentice Hall. ISBN: 0132827409, 9780132827409. 579pp.

Miller G. Tyler (2000). *Living in the Environment: Principles, Connections and Solutions.* 11th Edition. Brooks / Cole. ISBN: 0534376088, 9780534376086. 815pp.

Nebel, B. J., & Wright, R. T. (1993). *Environmental science: the way the world works.* Prentice Hall Professional.

Maggs, R., Wahid, A., Shamsi, S. R. A., & Ashmore, M. R. (1995). Effects of ambient air pollution on wheat and rice yield in Pakistan. *Water, Air, and Soil Pollution*, 85(3), 1311-1316.

Mazhar, F., & Jamal, T. (2009). Temporal population growth of Lahore. *Journal of Scientific Research*, 39(1).

Sami, J., Ali, M., Shafiq, M., Isma, Y., (2004). Air Quality in Lahore City – A study of pollutant gaseous emissions at Lahore. *Act Sci.*, 14(2) 87-94.

Schwartz, J. (1994). Air pollution and daily mortality: a review and meta-analysis. *Environmental research*, 64(1), 36-52.

Shafiq, M., Santos, B., Perez, A., and Blank, C.: *Global climate data exchange and differences in approaches of policy makers, scientists and the community*, Proceedings of the Earth Observation Symposium of 62nd International Astronautical

Congress (IAC), Paper ID: IAC-11.B1.6.4, Cape Town, 03.10.2011 – 07.10.2011. doi: 10.13140/2.1.2539.8569.

Muhammad Shafiq (2001) *"A study of different air pollutants in the commercial and residential areas of Lahore as compared with the past year (1999-2000)"*, Department of Space Science, University of the Punjab, Lahore, Pakistan (MSc Thesis).

Shirazi, S. A. (2012). Temporal Analysis of Land Use and Land Cover Changes in Lahore-Pakistan. *Pakistan Vision*, 13(1).

Smith, K. R. (1993). Fuel combustion, air pollution exposure, and health: the situation in developing countries. *Annual Review of Energy and the Environment*, 18(1), 529-566.

Stern, A. C., Boubel, R. W., Turner, D. B., & Fox, D. L. (1984). *Fundamentals of air pollution*. 2nd Edition, Orlando. Academic Press. ISBN: 0-12-666580-X.

The Pakistan National Conservation Strategy (1999), Government of Pakistan

Thomas, W., Rühling, A., & Simon, H. (1984). Accumulation of airborne pollutants (PAH, chlorinated hydrocarbons, heavy metals) in various plant species and humus. *Environmental Pollution Series A, Ecological and Biological*, 36(4), 295-310.

Thurston, G. D., Ito, K., Kinney, P. L., & Lippmann, M. (1991). A multi-year study of air pollution and respiratory hospital admissions in three New York State metropolitan areas: results for 1988 and 1989 summers. *Journal of exposure analysis and environmental epidemiology*, 2(4), 429-450.

Wahid, A., Maggs, R. S. R. A., Shamsi, S. R. A., Bell, J. N. B., & Ashmore, M. R. (1995). Effects of air pollution on rice yield in the Pakistan Punjab. *Environmental Pollution*, 90(3), 323-329.

Wahid, A., Maggs, R. S. R. A., Shamsi, S. R. A., Bell, J. N. B., & Ashmore, M. R. (1995). Air pollution and its impacts on wheat yield in the Pakistan Punjab. *Environmental Pollution*, 88(2), 147-154.

Zhou, Y., & Levy, J. I. (2007). Factors influencing the spatial extent of mobile source air pollution impacts: a meta-analysis. *BMC Public Health*, 7(1), 89.

Annex-2:
Bio of Authors

Dr. Isma Younes

PhD in Geography from Department of Geography, University of the Punjab, Lahore, Pakistan

Assistant Professor at Department of Geography, University of the Punjab, Lahore, Pakistan

E-mail: ismayounes2004@hotmail.com

Corresponding Author:

Dr.rer.nat. Muhammad Shafiq

PhD in Meteorology from Institute of Meteorology, University of Innsbruck, Austria

Research Scientist at Space and Upper Atmosphere Research Commission (SUPARCO), Karachi, Pakistan.

E-mail: m.shafiq.meteo@gmail.com

Dr. Fouzia Shafiq

PhD in Botany from Department of Botany, University of Karachi, Karachi, Pakistan

Assistant Professor at Department of Biological Sciences, University of Sargodha, Sargodha, Pakistan.

E-mail: fouzia.m.shafiq@gmail.com

Left blank for pagination

Annex-3: Questionnaire

1) Are you affected by pollution?
    a) Yes
    b) No

2) Is pollution a major problem?
    a) Yes
    b) No

3) Do you feel government is taking serious actions?
    a) Yes
    b) No

4) Did you adopt any step to control pollution at your own?
    a) Yes
    b) No

5) Do you have any information about EPA?
    a) Yes
    b) No

6) Which age group is more affected
    a) Children
    b) Young
    c) Old

7) Do you know that pollution is affecting your health?
    a) Yes
    b) No

8) What is the major source of pollution in your opinion?
    a) Rickshaw
    b) Wagons
    c) Other automobiles

9) Have you faced any problem regarding air pollution?
    a) Eye irritation
    b) Headache
    c) Nausea
    d) Lung problem

10) In your opinion can we overcome this problem?
    a) Yes
    b) No